IMO Problems, Theorems, and Methods

Number Theory

Mathematical Olympiad Series

ISSN: 1793-8570

Series Editors: Lee Peng Yee *(Nanyang Technological University, Singapore)*
Xiong Bin *(East China Normal University, China)*

Published

Vol. 28 *IMO Problems, Theorems, and Methods: Number Theory*
by Bin Xiong (East China Normal University, China) &
Gengyu Zhang (East China Normal University, China)

Vol. 27 *IMO Problems, Theorems, and Methods: Geometry*
by Tianqi Lin (Fudan University Affiliated High School, China) &
Bin Xiong (East China Normal University, China)

Vol. 26 *IMO Problems, Theorems, and Methods: Combinatorics*
by Guangyu Xu (East China Normal University, China) &
Zhenhua Qu (East China Normal University, China)

Vol. 25 *IMO Problems, Theorems, and Methods: Algebra*
by Jinhua Chen (East China Normal University, China) &
Bin Xiong (East China Normal University, China)

Vol. 24 *Leningrad Mathematical Olympiads (1961–1991)*
by Dmitri Fomin

Vol. 23 *Solving Problems in Point Geometry:*
Insights and Strategies for Mathematical Olympiad and Competitions
by Jingzhong Zhang (Guangzhou University, China &
Chinese Academy of Sciences, China) &
Xicheng Peng (Central China Normal University, China)

Vol. 22 *Mathematical Olympiad in China (2021–2022):*
Problems and Solutions
editor-in-chief Bin Xiong (East China Normal University, China)

Vol. 21 *Problem Solving Methods and Strategies in High School*
Mathematical Competitions
edited by Bin Xiong (East China Normal University, China) &
Yijie He (East China Normal University, China)

The complete list of the published volumes in the series can be found at
http://www.worldscientific.com/series/mos

Vol. 28 | Mathematical Olympiad Series

IMO Problems, Theorems, and Methods

Number Theory

Authors

Gengyu Zhang
Bin Xiong
East China Normal University, China

Proofreader

Jiu Ding
School of Mathematics and Natural Sciences,
University of Southern Mississippi, USA

Copy Editors

Lingzhi Kong, Liyu Zhang, and Ming Ni
East China Normal University Press, China

East China Normal University Press

World Scientific

Published by

East China Normal University Press
3663 North Zhongshan Road
Shanghai 200062
China

and

World Scientific Publishing Co. Pte. Ltd.
5 Toh Tuck Link, Singapore 596224
USA office: 27 Warren Street, Suite 401-402, Hackensack, NJ 07601
UK office: 57 Shelton Street, Covent Garden, London WC2H 9HE

Library of Congress Control Number: 2025004966

British Library Cataloguing-in-Publication Data
A catalogue record for this book is available from the British Library.

Mathematical Olympiad Series — Vol. 28
IMO PROBLEMS, THEOREMS, AND METHODS
Number Theory

Copyright © 2026 by East China Normal University Press and
World Scientific Publishing Co. Pte. Ltd.

ISBN 978-981-98-0333-0 (hardcover)
ISBN 978-981-98-0690-4 (paperback)
ISBN 978-981-98-0334-7 (ebook for institutions)
ISBN 978-981-98-0335-4 (ebook for individuals)

For any available supplementary material, please visit
https://www.worldscientific.com/worldscibooks/10.1142/14100#t=suppl

Desk Editors: Nambirajan Karuppiah/Angeline Husni

Typeset by Stallion Press
Email: enquiries@stallionpress.com

Preface

It is generally believed that formal mathematics competitions began with a contest held in Hungary in 1894, an event that gradually garnered attention worldwide. People aptly liken mathematics competitions to "Mental Gymnastics." In 1934, the Soviet Union straightforwardly termed it the "Mathematical Olympiad," a designation that reflects the Olympic spirit of pursuing excellence in intellect more vividly than the previous term, "mathematics competition."

By 1959, the internationalization of mathematics competitions had matured, leading to the inception of the "International Mathematical Olympiad" (IMO). The first IMO was held in Brasov, Romania in 1959. As of 2023, the IMO has successfully been held 64 times, except for 1980 when it was not conducted.

The IMO is typically held in July each year, and the format has become standardized: the official competition spans two days, with contestants tackling three problems in 4.5 hours each day, each problem worth a maximum of 7 points, totaling 42 points. Each participating team consists of six contestants, accompanied by a Leader and a Deputy Leader. Approximately half of the contestants receive medals, with about $1/12$ of the contestants earning gold medals, $2/12$ receiving silver medals, and $3/12$ obtaining bronze medals.

The IMO is currently one of the most influential secondary school mathematics competitions worldwide. In recent years, over 100 countries and regions have participated in this event, including all major nations globally.

Problems for the IMO are submitted by the participating teams and then reviewed and selected by a problem selection committee organized by the host country. This committee narrows down the submissions to approximately 30 Shortlist problems, covering algebra, geometry, number theory, and combinatorics, with about seven to eight problems on each topic. These are then presented to the Jury Meeting, composed of team leaders, who discuss and vote to decide on the six problems that will constitute the official competition paper. The host country does not provide any problems.

This event has played a significant role in promoting the exchange of mathematical education among nations, enhancing the level of mathematical education, facilitating mutual learning and understanding among young students worldwide, stimulating a broad interest in mathematics among secondary school students, and identifying and nurturing mathematically gifted students.

The development over more than 60 years is the result of the collective efforts of mathematicians, organizers, and contestants, and is worthy of reflection and study. Particularly deserving of study are the evolution of the competition problems, the mathematical ideas, and methods involved. Indeed, several colleagues from the International Mathematical Olympiad Research Center at East China Normal University had envisioned research and publication before the 60th IMO. For this purpose, we initiated several seminars involving over 10 people. For special reasons, this work was delayed. Based on the mathematical domains covered by the IMO problems — algebra, geometry, number theory, and combinatorics — we planned to compile the work into four volumes, with the general title *IMO Problems, Theorems, and Methods*, to be included in the "IMO Study Series".

Each volume begins with an introduction that provides an overview of the IMO. Subsequent chapters introduce relevant foundational knowledge and methods, followed by a reclassification and organization of past IMO problems. For some problems, multiple solutions are provided, along with a difficulty analysis. It is worth noting that some problems do not fit neatly into a single topic, as they may involve both algebra and number theory, or algebra and combinatorics. We primarily categorize them based on the topic under which they were placed on the Shortlist.

The four volumes titled *IMO Problems, Theorems, and Methods* were conceived with an overall writing plan proposed by myself, with the authors collectively discussing and refining the plan. The majority of the initial drafts were completed by Jinhua Chen (Algebra), Tianqi Lin (Geometry), Gengyu Zhang (Number Theory), and Guangyu Xu (Combinatorics). The first three volumes were supplemented, consolidated, and finalized by myself, while the combinatorics volume was supplemented, consolidated, and finalized by Zhenhua Qu.

We extend our gratitude to the leaders and contestants of the Chinese IMO teams over the years, as some elegant solutions included in the book were contributed by them. During the compilation of this book, we consulted various domestic and international sources, which are too numerous to acknowledge individually here.

While the authors have diligently studied the IMO problems and provided thoughtful strategies and solutions, errors and inaccuracies may occur due to our limitations. We sincerely invite readers to offer corrections and feedback.

The translation of the algebra volume in this series was done by Jinhua Chen and Bin Xiong; the geometry volume was translated by Xinyuan Yang; the number theory volume was translated by Gengyu Zhang; and the combinatorics volume was translated by Zhenhua Qu and Jinhua Chen. Jiu Ding revised the translations of all four books.

Bin Xiong
June 2024

About the Authors

Gengyu Zhang, graduated from School of Mathematical Sciences of Peking University as an undergraduate and Columbia University as a graduate. During high school he won two gold medals in Chinese Mathematical Olympiad (CMO). After graduation, he has always maintained his interest and passion in mathematical competitions, conducted research on formulating and solving mathematical problems, and participated in a wide range of mathematical education activities in primary and secondary schools.

Bin Xiong is a professor and doctoral supervisor at the School of Mathematical Sciences, East China Normal University. He also serves as the director of the Shanghai Key Laboratory of Core Mathematics and Practice, and the International Mathematical Olympiad Research Center. Professor Xiong is an expert with the State Council Special Allowance, and has been honored with the Shanghai May 1 Labor Medal as well as the Shanghai Model of Teaching and Educating. He has published over 100 scholarly papers in renowned national and international journals and has authored or co-authored more than 150 books. Additionally, Professor Xiong has served as the leader and head coach of the Chinese IMO team more than 10 times, and received the prestigious Paul Erdös Award in 2018 for his contribution to the development of mathematics competitions at the national and international level.

Contents

Introduction to the IMO

The International Mathematical Olympiad (IMO), established in the year 1959, represents one of the foremost intellectual endeavors at the highest tier for youth on a global scale. Prior to 1959, numerous countries around the world had already initiated the organization of mathematics competitions, thereby laying the groundwork for the inception of the IMO.

In 1891, the renowned physicist and President of the Hungarian Academy of Sciences, Loránd Eötvös (also known as Roland Eötvös), founded the Hungarian Mathematical and Physical Society. In 1894, he assumed the position of Minister of Education, and under his enthusiastic support, the Hungarian Mathematical and Physical Society initiated secondary school mathematics competitions. This competition, also known as the Eötvös Competition, offered winners the Eötvös Prize and the opportunity to pursue higher education. Subsequently, the Eötvös Competition was not held during 1919–1921 and 1944–1946 due to the world political events. In 1947, under the leadership of János Surányi, the Eötvös Competition was reinstated and renamed the Kürschák Competition (named after József Kürschák). This competition has played a significant role in Hungary in nurturing numerous mathematicians and scientists, including Győző Zemplén, Lipót Fejér, Theodore von Kármán, Alfréd Haar, Dénes Kőnig, Marcel Riesz, Gábor Szegő, Tibor Radó, Edward Teller, and Tibor Szele. Interestingly, George Pólya also participated in the competition, but did not hand in his paper.

With the aim of identifying and nurturing mathematical talents, prominent mathematicians such as Boris Delaunay (also known as Delone), Grigorii Fikhtengol'ts, Dmitry Faddeev, and others organized the inaugural Leningrad Mathematical Olympiad (LMO) in 1934, under the initiative of

Boris Delaunay. Winners of this competition were granted the privilege of direct admission to the Mathematics Department of Leningrad State University without the need for entrance examinations. Following the example set by the LMO, in 1935, renowned mathematicians Pavel Aleksandrov and Andrey Kolmogorov, alongside the entire faculty of the Mathematics Department at Moscow State University, organized the first Moscow Mathematical Olympiad (MMO).

Subsequently, various regions throughout the Soviet Union started hosting their own Mathematical Olympiads, ultimately laying the foundation for the All-Russian Mathematical Olympiad, which was first conducted in 1961. In 1967, the responsibility for organizing All-Russian Mathematical Olympiad was assumed by the Ministry of Education of the Soviet Union, leading to a renaming of the All-Russian Mathematical Olympiad as the All-Soviet-Union Mathematical Olympiad.

In fact, almost all of the best mathematicians born in the Soviet Union after 1930 had participated in Mathematical Olympiads, usually achieving first prizes. This distinguished group includes Fields Medal awardees such as Sergei Novikov, Grigory Margulis, Vladimir Drinfeld, Maxim Kontsevich, Grigori Perelman, and Stanislav Smirnov. Although having claimed that he was never particularly interested in Mathematical Olympiads, Sergei Novikov did secure a second prize in the MMO when he was in eighth grade.

While the United States of America Mathematical Olympiad (USAMO) was first held in 1972, the United States had a longstanding tradition of organizing mathematics competitions prior to that. In 1921, William Lowell Putnam published an article in the Harvard Graduates' Magazine, proposing the idea of conducting a university-level mathematics team competition. Following his passing, the Putnam family established the William Lowell Putnam Intercollegiate Memorial Fund to support the organization of the William Lowell Putnam Mathematical Competition (Putnam Competition), administered by the Mathematical Association of America.

With the assistance of George David Birkhoff, the first Putnam Competition took place in 1938, and it has been held annually since then; the top five ranking participants are designated as Putnam Fellows. Due to wartime conditions, the competition was not held from 1943 to 1945. In 1946, George Pólya, Tibor Radó, and Irving Kaplansky (Putnam Fellow in 1938) formed the Putnam Competition Committee, thus reestablishing the competition, but the responsibility for administration was undertaken by Garrett Birkhoff, the son of George David Birkhoff, and his colleagues in the Harvard University Department of Mathematics.

Many participants in the Putnam Competition have gone on to become prominent mathematicians and scientists. John Milnor, David Mumford, Daniel Quillen, Paul Cohen, John G. Thompson, and Manjul Bhargava have been recipients of Fields Medal. Richard Feynman, Kenneth Geddes Wilson, Steven Weinberg, and Murray Gell-Mann have received Nobel Prize in Physics, while John Nash was awarded the Nobel Prize in Economic Sciences. Additionally, numerous Putnam Fellows have been elected as members of the National Academy of Sciences in the United States.

Building upon the foundation of existing mathematics competitions in many countries, particularly in Eastern European nations, Romania proposed in 1956 the organization of an international mathematics competition involving seven Eastern European countries. This proposal led to the inaugural IMO held in 1959.

The first IMO was held in Braşov, Romania, in 1959. As of 2023, the IMO has been successfully held 64 times, with the exception of the year 1980 when it was not held for specific reasons. Apart from the 61st IMO, which was postponed to September in 2020 due to the impact of the COVID-19 pandemic, the IMO typically takes place in July each year.

The IMO has emerged as the most influential secondary school mathematics competition at present. In recent years, the number of countries and regions participating in this event has exceeded 100.

1 Evolution of the IMO

The first IMO, held in 1959, saw the participation of only 52 contestants from seven countries, including Romania, Hungary, Czechoslovakia, Bulgaria, Poland, the Soviet Union, and the German Democratic Republic. Subsequently, new countries and regions gradually joined this prestigious event. By the 20th IMO, also hosted by Romania in 1978, approximately 20 countries and regions participated (with 21 in the 19th, 17 in the 20th, and 23 in the 21st). The number of participating contestants also reached 132. The historical participation trends are depicted in Figure 1.

As the influence of the IMO continued to expand, the number of participating countries and regions, as well as the number of contestants, grew rapidly. The most significant increase in the number of participating countries and regions occurred in the 34th IMO, which was held in Turkey in 1993. In comparison to the 33rd IMO held in Russia in 1992, the number increased by 17 countries and regions, reaching a total of 73, with 413 contestants. By the 40th IMO, which was still hosted by Romania in 1999,

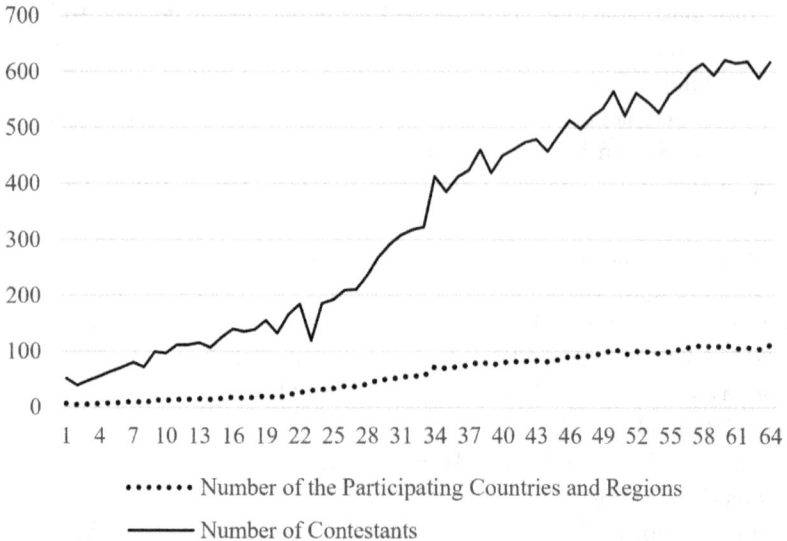

Figure 1 Numbers of Participating Countries and Regions, as Well as the Contestants, in the First 64 IMOs

the number of participating countries and regions had reached 81, with 450 contestants.

The first instance of the number of participating countries and regions surpassing 100 occurred during the 50th IMO, held in Germany in 2009, with a total of 565 contestants. Among the first 64 IMOs, the biggest number of participating countries and regions, as well as the largest number of contestants, was observed in the 60th IMO, hosted by the United Kingdom in 2019, where 621 contestants from 112 countries and regions took part. In the 64th IMO held in Japan in 2023, there were 618 contestants from 112 countries and regions.

As evident from Appendix A, the IMO is primarily hosted by European countries. Moreover, as the number of participating countries and regions in the IMO has increased, it is no longer confined to the seven founding member countries, and many new participating countries and regions have also begun to organize the IMO.

2 Problems in the IMO

The IMO is scheduled to take place annually in July. Each participating country or region officially sends a delegation consisting of six contestants, along with one team leader and one deputy leader. The official competition spans two days, with each day featuring three problems to be solved within a four-and-a-half-hour timeframe. Each problem carries a maximum score of 7 points, resulting in a total maximum score of 42 points, while the total maximum score of the team is 252 points.

In early IMOs, the number of problems and their individual point values varied from one session to another. For instance, the 2nd and 4th IMOs featured seven problems, while all other IMOs had 6 problems each. Additionally, in the 13th IMO, although the total score remained at 42 points, the six problems were allocated point values of 5, 7, 9, 6, 7, and 8, respectively. It was only from the 22nd IMO, held in the United States in 1981, that the IMO problems have become standardized, with each problem carrying 7 points and a total of six problems.

The number of contestants in each delegation has also become stable at six individuals starting from the 24th IMO held in France in 1983.

2.1 *The number of problems*

The mathematical domains covered by IMO problems encompass four major topics: algebra, combinatorics, geometry, and number theory. These are also the primary focus in various national mathematics competitions.

Across the 1st–64th IMOs, a total of 386 problems have been featured. Among them, geometry problems are the most numerous, with 123 problems, while number theory problems are the least, with 75 problems. Algebra problems account for approximately one-quarter of the total, comprising 101 problems. Furthermore, as indicated in Table 1, the quantity of algebra problems has remained relatively stable over each span of 10 IMOs.

Table 1 Numbers of Problems with Different Topics in the First 64 IMOs

Session	Topic			
	Algebra	Combinatorics	Geometry	Number Theory
1st–10th	20	6	29	7
11th–20th	20	12	18	10
21st–30th	14	16	18	12
31st–40th	13	16	15	16
41st–50th	15	11	20	14
51st–60th	13	19	18	11
61st–64th	6	7	6	5
Total	101	87	123	75

Remarkably, in the first 64 IMOs, there were two sessions when three number theory problems were presented, specifically in the 35th, 39th IMOs. In 14 IMOs, two number theory problems were featured, while in 41 IMOs, only one number theory problem was included. In seven IMOs, number theory problems were absent, namely the the 3rd, 5th, 7th, 8th, 15th, 18th, and 34th IMOs.

2.2   *The difficulty level of problems*

Typically, the difficulty of a problem is correlated with its problem number in the IMO.

Starting from the 24th IMO, the point value of each problem and the number of contestants per team have become standardized. Therefore, an analysis of the average scores of the 246 problems in the 24th–64th IMOs is presented in Table 2. It can be observed that the first and fourth problems in each IMO are relatively easy, with average scores generally exceeding 3 points. The second and fifth problems are relatively challenging, with average scores mainly ranging from 1 to 4 points. The third and sixth problems are exceptionally difficult, with average scores generally falling below 2 points.

The 246 problems are categorized into four topics: algebra, combinatorics, geometry, and number theory, as shown in Table 3. Notably, there is a relatively large representation of combinatorics and geometry problems. Combining this information with Table 1, it is evident that in the first 23 IMOs, algebra and geometry problems were predominant.

Furthermore, in early IMOs, geometry problems predominantly appeared in the 1st/4th and 2nd/5th positions. However, starting from

Table 2 Numbers of Problems with Different Average Scores in the 24–64 IMOs

	Problem Mean				
Problem Number	**0–1**	**1–2**	**2–3**	**3–4**	**4–7**
Problem 1	0	0	5	13	23
Problem 2	2	6	18	9	6
Problem 3	22	12	4	3	0
Problem 4	0	0	7	17	17
Problem 5	4	10	14	10	3
Problem 6	24	11	6	0	0
Total	52	39	54	52	49

Table 3 Numbers of Problems with Different Topics in the 24th to 64th IMOs

	Algebra			Combinatorics			Geometry			Number Theory		
						Topic						
Session	1st/4th	2nd/5th	3rd/6th	1st/4th	2nd/5th	3rd/6th	1st/4th	2nd/5th	3rd/6th	1st/4th	2nd/5th	3rd/6th
24th–30th	4	1	5	4	3	5	5	7	0	1	3	4
31st–40th	4	5	4	5	4	7	6	9	0	5	2	9
41st–50th	3	8	4	3	2	6	10	6	5	5	4	5
51st–60th	5	6	2	4	7	8	7	3	6	3	4	4
61st–64th	1	3	2	2	2	3	3	1	2	2	2	1
Total	17	23	17	18	18	29	31	26	13	16	15	23
		57			65			70			54	

the 41st–50th IMOs, geometry problems were more commonly found in the 1st/4th and 3rd/6th positions. Similarly, algebra problems were more frequent in the 1st/4th and 2nd/5th positions, combinatorics problems were more prevalent in the 2nd/5th and 3rd/6th positions, while the quantity of number theory problems across different problem numbers does not differ significantly.

From Table 4, it can be observed that among the four topics, the numbers of problems with an average score ranging from 2 to 4 points are quite similar. However, in the combinatorics topic, there is a higher quantity of challenging problems, with 31 problems having an average score between 0 and 2 points. Conversely, the geometry topic has the largest number of relatively easy problems, with 23 problems scoring above 4 points. This discrepancy is largely due to the fact that there are 29 combinatorics problems in the 3rd/6th positions, and 31 geometry problems in the 1st/4th positions.

Table 4 Numbers of Problems with Different Average Scores by Topic in the 24th–64th IMOs

Topic	Problem Mean					Total
	0–1	1–2	2–3	3–4	4–7	
Algebra	7	14	19	6	11	57
Combinatorics	20	11	12	14	8	65
Geometry	13	5	14	15	23	70
Number theory	12	9	9	17	7	54
Total	52	39	54	52	49	246

Furthermore, when considering Tables 3 and 4, it becomes apparent that among the four topics, the number of problems with an average score ranging from 0 to 2 points closely aligns with the number of problems in the 3rd/6th positions. There are slightly more problems with an average score between 2 and 4 points compared to those in the 2nd/5th positions, and slightly fewer problems with an average score exceeding 4 points compared to those in the 1st/4th positions. This indicates that even the seemingly easier problems in the IMO are not as straightforward as they might appear.

Notably, among these 246 problems, the lowest average score is attributed to IMO 58-3 (Combinatorics, proposed by Austria):

A hunter and an invisible rabbit play a game in a Euclidean plane. The rabbit's starting point, A_0, and the hunter's starting point, B_0, are the same. After $n-1$ rounds of the game, the rabbit is at point A_{n-1} and the hunter is at point B_{n-1}. In the nth round of the game, three things occur in order.

(i) The rabbit moves invisibly to a point A_n such that the distance between A_{n-1} and A_n is exactly 1.

(ii) A tracking device reports a point P_n to the hunter. The only guarantee provided by the tracking device to the hunter is that the distance between P_n and A_n is at most 1.

(iii) The hunter moves visibly to a point B_n such that the distance between B_{n-1} and B_n is exactly 1.

Is it always possible, no matter how the rabbit moves, and no matter what points are reported by the tracking device, for the hunter to choose her moves so that after 10^9 rounds she can ensure that the distance between her and the rabbit is at most 100?

This unconventional problem received an average score of only 0.042 points. Only two contestants, Mikhail Ivanov from Russia and Linus Cooper from Australia, achieved a perfect score of 7 points. Joe Benton from the United Kingdom scored 5 points, Pavel Hudec from Czech Republic earned 4 points, Hadyn Ka Ming Tang from Australia, Yahor Dubovik from Belarus, and Jeonghyun Ahn from South Korea each scored 1 point.

Furthermore, among the 20 lowest-scoring problems in the 24th to 64th IMOs, nearly all of them appeared in the 3rd/6th positions, as indicated in Table 5. There were three algebra problems, eight combinatorics problems,

Table 5 The 20 Problems with the Lowest Average Scores in the 24th–64th IMOs

Problem	Mean	Topic	Problem	Mean	Topic
IMO 58-3	0.042	Combinatorics	IMO 54-6	0.296	Combinatorics
IMO 48-6	0.152	Algebra	IMO 48-3	0.304	Combinatorics
IMO 50-6	0.168	Combinatorics	IMO 52-6	0.318	Geometry
IMO 47-6	0.187	Geometry	IMO 53-6	0.336	Number Theory
IMO 57-3	0.251	Number Theory	IMO 55-6	0.339	Combinatorics
IMO 49-6	0.260	Geometry	IMO 56-6	0.355	Combinatorics
IMO 64-6	0.275	Geometry	IMO 51-6	0.368	Algebra
IMO 59-3	0.278	Combinatorics	IMO 62-3	0.372	Geometry
IMO 61-6	0.282	Combinatorics	IMO 62-2	0.375	Algebra
IMO 58-6	0.294	Number Theory	IMO 60-6	0.403	Geometry

six geometry problems, and three number theory problems among them. These three number theory problems were:

- **(IMO 57-3, proposed by Russia)** Let $P = A_1 A_2 \cdots A_k$ be a convex polygon in a plane. The vertices $A_1, A_2, \ldots, A_k$ have integral coordinates and lie on a circle. Let S be the area of P. An odd positive integer n is given such that the squares of the side lengths of P are integers divisible by n. Prove that $2S$ is an integer divisible by n.

- **(IMO 58-6, proposed by the United States)** An ordered pair (x, y) of integers is a primitive point if the greatest common divisor of x and y is 1. Given a finite set S of primitive points, prove that there exist a positive integer n and integers $a_0, a_1, \ldots, a_n$ such that, for each (x, y) in S,

$$a_0 x^n + a_1 x^{n-1} y + a_2 x^{n-2} y^2 + \cdots + a_{n-1} xy^{n-1} + a_n y^n = 1.$$

- **(IMO 53-6, proposed by Serbia)** Find all positive integers n for which there exist non-negative integers $a_1, a_2, \ldots, a_n$ such that

$$\frac{1}{2^{a_1}} + \frac{1}{2^{a_2}} + \cdots + \frac{1}{2^{a_n}} = \frac{1}{3^{a_1}} + \frac{2}{3^{a_2}} + \cdots + \frac{n}{3^{a_n}} = 1.$$

2.3 The classification of problems

In the 1st–64th IMOs, there were 75 number theory problems, which can be categorized into three specialized subjects: divisibility of integers, modular arithmetic, and indeterminate equations, as shown in Table 6.

Table 6 Numbers of Number Theory Problems in the First 64 IMOs

Session	Subject		
	Divisibility	Modular Arithmetic	Indeterminate Equations
1st–10th	4	1	2
11th–20th	4	5	1
21st–30th	5	3	4
31st–40th	11	4	1
41st–50th	4	8	2
51st–60th	5	3	3
61st–64th	3	1	1
Total	36	25	14

In the first 64 IMOs, there had been a total of 36 divisibility problems, accounting for approximately 48.0% of all number theory problems. These problems can be primarily categorized into four types: (1) discussions on divisibility, totaling 14 problems; (2) prime numbers, prime factors, and coprime numbers, totaling eight problems; (3) function related problems, totaling five problems; (4) other problems, totaling eight problems.

As shown in Table 7, in the 24th–64th IMOs, there were 26 divisibility problems. These problems were predominantly found in the 3rd/6th positions, with a relatively balanced distribution among different problem types.

Table 7 Numbers of Divisibility Problems in the 24th–64th IMOs

Divisibility Problem	Problem Number			Number of Problems in the First 64 IMOs
	1, 4	2, 5	3, 6	
Discussions on divisibility	6	1	2	15
Prime numbers, prime factors and coprime numbers	0	3	4	8
Function related problems	0	1	4	5
Other problems	1	1	3	8
Total	7	6	13	36

In the first 64 IMOs, there had been a total of 25 modular arithmetic problems, accounting for approximately 33.3% of all number theory problems. These problems can be primarily categorized into four types: (1) existence problems, totaling 10 problems; (2) finding numbers that satisfy given conditions, totaling eight problems; (3) exploring relationships between terms, totaling four problems; (4) maximum or minimum value problems, totaling three problems.

As shown in Table 8, in the 24th–64th IMOs, there were 19 modular arithmetic problems. These problems were also predominantly found in the 3rd/6th positions, with the main focus being on existence problems and finding numbers that satisfy given conditions. The other two types of modular arithmetic problems appeared less frequently in the last 40 IMOs.

In the first 64 IMOs, there had been a total of 14 indeterminate equation problems, accounting for approximately 18.7% of all number theory problems. These problems can be primarily categorized into two types: (1) finding solutions of indeterminate equations, totaling 10 problems; (2) proving properties satisfied by indeterminate equations, totaling four problems.

As shown in Table 9, in the 24th–64th IMOs, there were nine indeterminate equation problems. These problems were predominantly present in

Table 8   Numbers of Modular Arithmetic Problems in the 24th–64th IMOs

Modular Arithemetic Problem	Problem Number			Number of Problems in the First 64 IMOs
	1, 4	2, 5	3, 6	
Existence problems	0	2	5	10
Finding numbers	2	1	4	8
Exploring relationships	1	2	0	4
Maximum or minimum values	2	0	0	3
Total	5	5	9	25

Table 9   Numbers of Indeterminate Equation Problems in the 24th–64th IMOs

Indeterminate Equation Problem	Problem Number			Number of Problems in the First 64 IMOs
	1, 4	2, 5	3, 6	
Finding solutions	2	4	0	10
Proving properties	2	0	1	4
Total	4	4	1	14

the 1st/4th positions as well as in the 2nd/5th positions, with the main focus being on finding solutions of indeterminate equations.

We can see that the number of problems for all the three types of number theory problems are almost the same in the 1st/4th and 2nd/5th positions. Overall, there are more problems in the 3rd/6th positions.

As shown in Table 10, there are 21 and 26 problems with average scores of 0–2 and 2–4 points, respectively, which are significantly more than those with average scores above 4 points. Comparing different positions of these

Table 10   Numbers of Number Theory Problems with Different Average Scores in the 24th–64th IMOs

Number Theory Problem	Problem Mean					Total
	0–1	1–2	2–3	3–4	4–7	
Divisibility of integers	6	3	5	8	4	26
Modular arithmetic	5	5	3	5	1	19
Indeterminate equations	1	1	1	4	2	9
Total	12	9	9	17	7	54
	21		26		7	

problems, we can see that even the easier number theory problems are not necessarily easy.

Finally, most number theory problems are comprehensive. For example, when solving indeterminate equations, properties related to divisibility of integers and congruence are commonly used. Sometimes, it is difficult to perfectly classify a number theory problem into one of the three categories above. For these problems, we classify them mainly based on what tools are used in solving them.

2.4 *The proposal for problems*

Problems in the IMO are proposed by the participating countries and regions, except the host. Usually the team leaders are in charge of submitting problems with a limit of six, and these problems are then subjected to selection by a selection committee composed of experts organized by the host. Approximately 30 problems are chosen as shortlist problems, with around eight problems in each of the four topics: algebra, geometry, combinatorics, and number theory. Subsequently, these problems are submitted to the Jury, which is comprised of team leaders from each participating country or region. The problems are discussed and voted upon to select the official examination problems. Once the problems are finalized, they are translated into five working languages: English, French, German, Russian, and Spanish. Each leader then translates the problems into their respective national languages, and contestants can choose from two languages in which to answer the problems.

Among the first 64 IMOs, number theory problems were contributed by 31 different countries and regions. Romania and the United Kingdom had the highest number of problems proposed, with a total of eight, followed by Poland with six problems. The Netherlands, Australia, Bulgaria, Germany, and the United States each proposed four problems, while Russia contributed three problems. These nine countries collectively provided 45 problems. Remarkably, only Romania (in the 43th IMO in 2002) have proposed two number theory problems in the same IMO session.

Furthermore, as indicated in Appendix B, number theory problems in the first 15 IMOs were primarily proposed by the seven founding countries of the IMO. From the 16th to 35th IMOs, number theory problems were predominantly contributed by European countries and regions. However, in the 36th–64th IMOs, number theory problems demonstrated a more diverse range of proposing countries and regions. This, to some extent, correlates with the expansion of the IMO's influence and the growth in the number of participating countries and regions.

3   Awards in the IMO

In addition to selecting problems, the Jury has several other responsibilities, including: establishing grading criteria, resolving discrepancies in grading between leaders and coordinators, and determining the number of gold, silver, and bronze medals, as well as the score thresholds. In each IMO, approximately 1/12 of the contestants receive a gold medal, 2/12 receive silver, and 3/12 receive bronze.

Apart from the gold, silver, and bronze medals, contestants who do not receive medals but attain a score of 7 on at least one problem in the IMO will receive an Honorable Mention. Contestants who deliver exceptionally elegant solutions to specific problems in the IMO will receive a Special Prize.

As depicted in Figure 2, starting from the 24th IMO, the cutoff scores for gold, silver, and bronze medals have gradually stabilized. The gold medal cutoff is approximately 29 points, the silver medal cutoff is around 22 points, and the bronze medal cutoff is roughly 15 points. Furthermore, the average

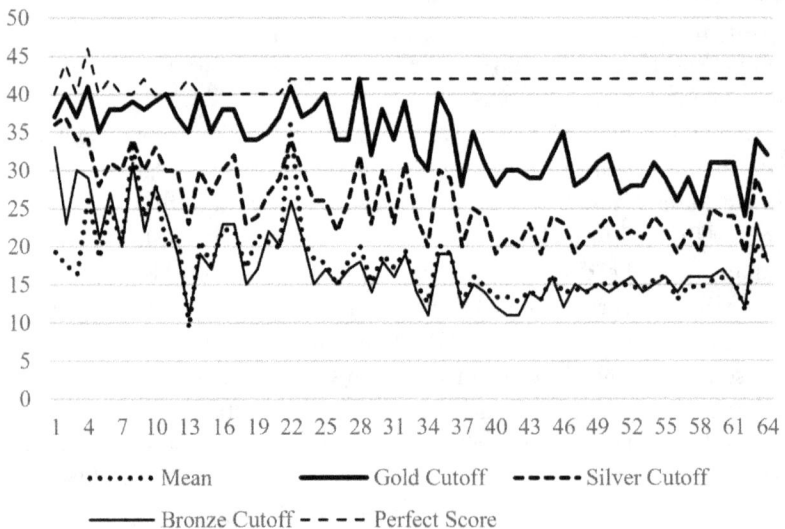

Figure 2   Medal Cutoff Scores in the First 64 IMOs

score of all contestants closely aligns with the bronze medal cutoff. This indicates that the problem difficulty is well-balanced.

Interestingly, in the first 64 IMOs, there were three occasions where the gold medal cutoff was a perfect score, meaning only those who scored full marks could earn a gold medal. These three IMOs were: the 11th IMO (1969, Romania) with a perfect score of 40 points and three gold medalists; the 14th IMO (1972, Poland) with a perfect score of 40 points and eight gold medalists; and the 28th IMO (1988, Cuba) with a perfect score of 42 points and 22 gold medalists.

3.1 *Participation*

In the first 64 IMOs, a total of 269 contestants took part in four or more IMOs. Among them, two contestants attended seven IMOs, four contestants attended six IMOs, 42 contestants attended five IMOs, and 221 contestants attended four IMOs (Table 11).

Table 11 Contestants with Six or More Participations in the First 64 IMOs

Contestant	Country	Participation Year	Gold	Silver	Bronze	Honorable Mention	Perfect Score
David Kunszenti-Kovács	Norway	1997–2003	1	3	1	1	
Yeoh Zi Song	Malaysia	2014–2020	1	1	4	1	
Zhuo Qun (Alex) Song	Canada	2010–2015	5	0	1	0	1
Teodor von Burg	Serbia	2007–2012	4	1	1	0	
Alexey Entin	Israel	2000–2005	1	3	1	0	
Tan Li Xuan	Malaysia	2016–2021	0	2	2	1	

Coincidentally, in the 43rd IMO held in 2002, the gold medal cutoff was set at 29 points, and David Kunszenti-Kovács achieved a total score of exactly 29 points.

In the 44th IMO held in 2003, the gold medal cutoff was set at 29 points, whereas Alexey Entin attained a total score of exactly 28 points.

In the 51st IMO held in 2010, the bronze medal cutoff was set at 15 points, and Zhuo Qun (Alex) Song achieved a total score of exactly 15 points.

In the 58th IMO held in 2017, the silver medal cutoff was set at 19 points, and Yeoh Zi Song achieved a total score of exactly 19 points. In the 59th IMO held in 2018, the silver medal cutoff was set at 25 points, and Yeoh Zi Song's total score was exactly 24 points. In the 61st IMO held in

2020, the gold medal cutoff was set at 31 points, and Yeoh Zi Song's total score was exactly 31 points.

Additionally, from 2002 to 2005 and in 2007, Sherry Gong participated in the IMO, earning one bronze, two silver, and one gold medal. In the 48th IMO held in 2007, she ranked 7th individually. Notably, from 2002 to 2004, she was a member of the Puerto Rico IMO team, while in 2005 and 2007, she was a member of the United States IMO team.

From 2020 to 2023, Alex Chui participated in the IMO, securing two gold and two silver medals. However, in 2020 and 2021, he was a member of the Chinese Hong Kong IMO team, while in 2022 and 2023, he was a member of the United Kingdom IMO team.

Other than Sherry Gong and Alex Chui, the remaining 267 contestants hailed from 75 different countries and regions. Among them, there were 12 contestants from Cyprus and Moldova each, 11 from Malaysia, eight from Trinidad and Tobago, seven from each of Estonia, Germany, Sri Lanka, and North Macedonia, and six from Japan and Philippines each.

3.2   *Gold medals*

In the first 64 IMOs, a total of 49 contestants achieved three or more gold medals. Among them, one contestant earned five gold medals, six contestants earned four gold medals, and 42 contestants earned three gold medals.

As shown in Table 12, the data source is from the IMO official website. Zhuo Qun (Alex) Song, Reid Barton, and Lisa Sauermann have all achieved perfect scores.

Table 12   Contestants with Four or More Gold Medals in the First 64 IMOs

Contestant	Country	Participation Year	Gold Year	Perfect Score Year
Zhuo Qun (Alex) Song	Canada	2010–2015	2011–2015	2015
Reid Barton	The United States of America	1998–2001	1998–2001	2001
Christian Reiher	Germany	1999–2003	2000–2003	
Lisa Sauermann	Germany	2007–2011	2008–2011	2011
Teodor von Burg	Serbia	2007–2012	2009–2012	
Nipun Pitimanaaree	Thailand	2009–2013	2010–2013	
Luke Robitaille	The United States of America	2019–2022	2019–2022	

Coincidentally, in the 54th IMO held in 2013, the gold medal cutoff was 31 points, and Nipun Pitimanaaree achieved a total score of exactly 31 points.

In the 41st IMO held in 2000, the top four contestants all achieved perfect scores, and Reid Barton ranked fifth with a total score of 39. In the 43rd IMO held in 2002, the top three contestants all achieved perfect scores, and Christian Reiher ranked fourth with a total score of 36. Furthermore, in the 50th IMO held in 2009, Lisa Sauermann achieved a total score of 41, securing the third position.

Moreover, Oleg Golberg participated in the IMO from 2002 to 2004, achieving three gold medals. He consistently ranked within the top 10 in terms of total scores. Notably, in 2002 and 2003, he was a member of the Russia IMO team, and in 2004, he was a member of the United States IMO team.

Apart from Oleg Golberg, the remaining 48 contestants hailed from 22 different countries and regions. Among them, there were four contestants from each of Russia, Bulgaria, Germany, Hungary, Romania, and the United States. Both South Korea and the United Kingdom had three contestants, while Canada, Japan, Singapore, and the Soviet Union were represented by two contestants each.

3.3 *Special prizes*

In the first 64 IMOs, only 44 contestants have received special prizes. Among them, one contestant has received the special prize three times, seven contestants have earned twice, and 36 contestants have achieved once. It indicates that achieving a special prize is even more challenging than securing a gold medal.

As shown in Table 13, the data source is from the IMO official website. John Rickard, Imre Ruzsa, and Marc van Leeuwen all achieved two special prizes in one IMO for their elegant solutions. Furthermore, John Rickard, Imre Ruzsa, and László Lovász have all earned a perfect score twice.

Coincidentally, in the 11th IMO held in 1969, only three gold medals were awarded, with Imre Ruzsa ranking 4th and receiving a silver medal. Similarly, in the 19th IMO held in 1977, which also resulted in only 13 gold medals, Marc van Leeuwen ranked 14th and earned a silver medal.

Additionally, these 44 contestants hailed from 16 different countries and regions. Among them, there were seven contestants from each of Hungary and the United Kingdom, five from the German Democratic Republic, four

Table 13 Contestants with Multiple Special Prizes in the First 64 IMOs

Contestant	Country	Participation Year	Special Prize Year	Gold Year	Perfect Score Year
John Rickard	The United Kingdom	1975–1977	1976, 1977 (2)	1975–1977	1975, 1977
József Pelikán	Hungary	1963–1966	1965, 1966	1964–1966	1966
László Lovász	Hungary	1963–1966	1965, 1966	1964–1966	1965, 1966
László Babai	Hungary	1966–1968	1966, 1968	1968	1968
Simon Phillips Norton	The United Kingdom	1967–1969	1967, 1969	1967–1969	1969
Wolfgang Burmeister	The German Democratic Republic	1967–1971	1970, 1971	1968, 1970, 1971	1970
Imre Ruzsa	Hungary	1969–1971	1971 (2)	1970, 1971	1970, 1971
Marc van Leeuwen	The Netherlands	1977, 1978	1978 (2)		

from Bulgaria and Poland each, three from each of Czechoslovakia and the Soviet Union, and two from Finland and the United States each.

Special prizes were more frequently granted in the first 20 IMOs, with a total of 27 special prizes earned by 22 contestants from the 11th to 20th IMOs. Subsequently, the frequency of special prize presentations declined. Since Moldovan contestant Iurie Boreico received a special prize for his brilliant solution to IMO 46-3 (Algebra, proposed by South Korea) in 2005, no contestant has achieved this accolade to date.

From 2003 to 2007, Iurie Boreico consistently participated in the IMO, earning three gold and two silver medals. He achieved a perfect score in 2005 and 2006. It's noteworthy that the 44th IMO held in 2003 only yielded 37 gold medals, with Iurie Boreico placing 38th individually and receiving a silver medal.

- **(IMO 46-3, proposed by South Korea).** Let x, y, z be three positive reals such that $xyz \geq 1$. Prove that

$$\frac{x^5 - x^2}{x^5 + y^2 + z^2} + \frac{y^5 - y^2}{y^5 + z^2 + x^2} + \frac{z^5 - z^2}{z^5 + x^2 + y^2} \geq 0.$$

4 Summary

The IMO stands as a distinguished intellectual competition for young minds. According to a study by Agarwal R. and Gaule P., statistical analysis reveals that among contestants in the IMO (including those who did not secure medals), 22% choose to pursue further studies in mathematics, ultimately obtaining doctoral degrees in the field. Additionally, 1% of these contestants become presenters at the International Congress of Mathematicians, and 0.2% attain the Fields Medal. These statistics underscore the vital role of the IMO in identifying and nurturing mathematical talent.

It's essential not to perceive the IMO as a mere selection exam. Rather than focusing solely on the brief two-day competition, the crucial aspect lies in the learning and preparation undertaken before participating. As the mathematician Paul Halmos aptly put it, what mathematics really consists of is problems and solutions. Contestants, through their exploration of Olympiad problems, not only enhance their mathematical abilities but also experience the joy and satisfaction of problem-solving. This experience plants the seeds of a future career in mathematics.

However, it's important to acknowledge that Olympiad problems and research problems in mathematics differ. Research problems often lack readily available answers and may require the investment of countless days and nights. Hence, the IMO is just one pathway in the growth of mathematical talents, and success in the IMO is not the sole qualification for becoming an outstanding mathematician.

Although every contestant aims for a gold medal, their aspirations go far beyond accolades. On this stage, they have the opportunity to showcase their intellectual capabilities, revel in the mathematical exploration, and relish competing with talented young minds from around the world, all without the narrow goal of proving their superiority over others. While the competition results may vary, each contestant stands as a victor in their own right and becomes a companion and witness to one another's life journeys.

In contrast to the Olympics, where athletes' careers are closely intertwined with the Games, the IMO is merely a chapter in the growth of these gifted young individuals. Following the IMO, the door to a new mathematical world has already swung wide open for them.

Chapter 1

Divisibility of Integers

For most people, integers (especially positive integers) are probably one of the first concepts they learn about mathematics. Through the four fundamental operations of integers (addition, subtraction, multiplication and division), we glimpse into the mysterious realm of mathematics. From simply counting numbers, to one of the most famous propositions — also known as the pearl on the crown of mathematics — the Goldbach conjecture, integers and the number theory system are closely related to so many facets of mathematics.

In high school mathematics competitions, one major category, "number theory," primarily focuses on problems related to integers. From the 1st to the 64th IMO, there are 75 number theory problems in total, covering various aspects of elementary number theory. In the rest of this book, we will categorize number theory problems into three major classes: divisibility of integers, modular arithmetic, and indeterminate equations. It is worth mentioning that number theory itself is not fragmented; it is not feasible to strictly differentiate all problems according to the above classifications. At the same time, within the entire Mathematical Olympiad system, number theory is not detached; many number theory problems actually incorporate a considerable portion of algebra, combinatorics, or even geometry, such as functional equations, the pigeonhole principle and so on. Therefore, our categorization has a certain degree of subjectivity. On the one hand, it aims to reflect some characteristics of different number theory problems; on the other hand, it is intended to prevent the entire number theory part from becoming overly lengthy, ensuring the readability for readers. For those problems combined with knowledge other than number theory, if they have no concern with congruence or indeterminate equations, then they will be classified into the "Divisibility of Integers" section.

The discussion on divisibility of integers has a long history. Conventionally, we denote the set of integers, or the ring of integers, as **Z**. This notation originates from the German female mathematician Emmy Noether, who made significant contributions to the study of rings. Since the German word for "counting numbers" is "Zählen," she used **Z** to denote the ring of integers, and this notation has been passed down to the present day. In fact, as a type of integral domain in advanced mathematics, one of the crucial characteristics of the set of integers is that it allows the definition of addition, subtraction, and multiplication, and the set of integers is closed for these three operations. Therefore, one of our primary concerns regarding integers is the result of division operations, that is, whether the result of one integer dividing another integer is still an integer. This is also the most common type of number theory problems in Olympiad — the divisibility of integers.

In the first 64 IMOs, there had been a total of 36 divisibility problems, accounting for approximately 48.0% of all number theory problems. These problems can be primarily categorized into four types: (1) discussions on divisibility, totaling 15 problems; (2) prime numbers, prime factors, and coprime numbers, totaling eight problems; (3) function related problems, totaling five problems; (4) other problems, totaling eight problems. The statistical distribution of these four types of problems in previous IMOs is presented in Table 1.1.

It can be observed that divisibility of integers is a focal point of number theory problems. Except for the 41st to the 50th IMO, divisibility problems accounted for about half of all number theory problems in each session of 10 IMOs. In the 31st to the 40th IMO, the proportion even approached 70%.

Furthermore, in the first 50 IMOs, the number of problems related to discussion on divisibility remained relatively stable, with approximately 3 such problems appearing in every 10 IMOs. However, these days there are various methods for problems simply related to divisibility. Therefore, such problems are relatively straightforward and in recent years these problems have not been prevalent. Instead, there has been an emphasis on more specific issues such as prime numbers, prime factors, coprime numbers, or other problems with strong integrative aspects.

Table 1.1 Numbers of Divisibility Problems in the First 64 IMOs

Content	Sessions							Total
	1–10	11–20	21–30	31–40	41–50	51–60	61–64	
Discussions on divisibility	2	2	4	3	3	0	1	15
Prime numbers, prime factors, and coprime numbers	0	1	1	3	1	1	1	8
Fuction related problems	0	0	0	3	0	2	0	5
Other problems	2	1	0	2	0	2	1	8
Number theory problems	7	10	12	16	14	11	5	75
The percentage of divisibility problems among number theory problems	57.1%	40.0%	41.7%	68.8%	28.6%	45.5%	60.0%	48.0%

Many number theory problems in IMOs combined algebraic, combinatorial, and even geometric methods. These problems were more comprehensive, and contestants usually scored less points on these problems. For instance, function related problems are those problems discussing divisibility properties within functional equations.

This chapter will be divided into three parts. The first part introduces some properties and theorems related to divisibility, which are not very difficult. Readers can find more details of applications in the second part, through the discussion on IMO problems and solutions.

The second part revolves around four types of problems: "discussions on divisibility," "prime numbers, prime factors, and coprime numbers," "function related problems," and "other problems." These problems are presented in chronological order, and some problems include various solutions and generalizations.

It is important to note that for each problem, the solutions are followed by information on the scores, including the number of contestants in each score range, the average score, and the scores of the top five teams. However, early IMOs often lacked information on contestant scores, so the number of contestants in each score range only represents the counted number of contestants, and some problems lack scores of the top five teams.

The third part provides a brief summary of this chapter.

1.1   Common Properties, Theorems, and Methods

1.1.1   *Definition and properties of divisibility*

(1) *Definition of divisibility*

For any integers a, b ($a \neq 0$), if there exists an integer q such that $b = aq$, then b is divisible by a, or we say a divides b. This is written as $a \mid b$. Otherwise, we define b as not divisible by a, or a does not divide b, written as $a \nmid b$.

(2) *Transitivity of divisibility*

If $a \mid b$ and $b \mid c$, then $a \mid c$.

(3) *Definition of primes*

For any positive integer $n > 1$, if n has no positive divisors except 1 and n itself, then n is a prime number (or a prime); otherwise, we call n a composite number.

(4) *Properties of primes*

  (i) The smallest prime number is 2, and 2 is the only prime number which is even.

 (ii) There are infinitely many prime numbers.

 (ii) Suppose n is a positive integer greater than 1. If n is not divisible by any prime number p which is greater than 1 and less than $\sqrt{n}$, then n is a prime number.

(5) *Greatest common divisor*

Suppose a and b are two integers which are not both zero, d is a non-zero integer. If $d \mid a$ and $d \mid b$, then d is a common divisor of a and b. It is trivial that the set of common divisors of a and b has a maximum, which is defined as the greatest common divisor (GCD), or the highest common factor (HCF) of a and b, denoted as (a, b). If $(a, b) = 1$, then we say a is coprime to b, or a is relatively prime to b.

(6) *Least common multiple*

Suppose a, b, and c are three non-zero integers. If $a \mid c$ and $b \mid c$, then c is a common multiple of a and b. Among all the positive multiples of a and b, the least one is defined as the least common multiple of a and b, denoted as $[a, b]$.

(7) *The Euclidean algorithm*

For non-zero integers a and b with $|a| \le |b|$, we can use the Euclidean algorithm to compute the greatest common divisor of a and b.

By division, let $b = aq_1 + r_1$, where q_1, r_1 are integers and $0 \le r_1 < |a|$. If $r_1 = 0$, then b is divisible by a. Otherwise we divide a by r_1, denoted as $a = r_1 q + r_2$, $0 \le r_2 < r_1$. Repeating this process gives successively $r_1 > r_2 > \cdots$, until some $r_{k+1} = 0$. Now the greatest common divisor of a and b is r_k.

1.1.2 *Common theorems and methods for divisibility*

(1) *Bézout's theorem*

Suppose $d = (a, b)$. Then there exist integers x and y such that $ax + by = d$. This theorem can be proved by the Euclidean algorithm.

Moreover, if a is coprime to b, then there exist integers x and y such that $ax + by = 1$. The converse proposition is also true.

For more than two integers, if integers $a_1, a_2, \ldots, a_n$ are not all zero, then there exist integers $x_1, x_2, \ldots, x_n$, such that

$$a_1 x_1 + a_2 x_2 + \cdots + a_n x_n = (a_1, a_2, \ldots, a_n).$$

Example 1.1. Prove that for any positive integer n, there exist integers a and b, such that n divides $4a^2 + 9b^2 - 1$.

Proof. If n is odd, then we write it as $n = 2k + 1$, where k is a non-negative integer. Let $a = k, b = 0$, and then

$$4a^2 + 9b^2 - 1 = (2k + 1)(2k - 1)$$

is divisible by n.

If n is not divisible by 3, then we write it as $n = 3k + r$, where k is a non-negative integer, $r \in \{-1, 1\}$. Let $a = 0, b = k$, and then

$$4a^2 + 9b^2 - 1 = (3k + 1)(3k - 1)$$

is divisible by n.

Now we only need to consider the case when n is a multiple of 6. We write it as $n = 2^r 3^s m$, where r, s are positive integers, and m is relatively prime to 6. Since 2^r is relatively prime to $3^s m$, by Bézout's theorem, there exist non-zero integers k, l such that

$$2^r k + 3^s ml = 1,$$

which implies

$$2^{2r} k^2 + 3^{2s} m^2 l^2 + 2 \cdot 2^r 3^s mkl = 1,$$

$$-2nkl = 2^{2r} k^2 + 3^{2s} m^2 l^2 - 1.$$

Then let $a = 2^{r-1}k, b = 3^{s-1}ml$, and we see that

$$4a^2 + 9b^2 - 1 = 2^{2r} k^2 + 3^{2s} m^2 l^2 - 1$$

is divisible by n.

(2) *Fundamental theorem of arithmetic*

Every integer n greater than 1 can be represented uniquely as a product of prime numbers, up to the order of the factors. This theorem is also known as the unique factorization theorem or prime factorization theorem. The factorization of n can be presented as

$$n = p_1^{\alpha_1} p_2^{\alpha_2} \cdots p_k^{\alpha_k},$$

where $p_1, p_2, \ldots, p_k$ are distinct prime numbers, $\alpha_1, \alpha_2, \ldots, \alpha_k$ are positive integers.

(3) *The formula for calculating the number of positive factors*

Suppose the prime factorization of n is

$$n = p_1^{\alpha_1} p_2^{\alpha_2} \cdots p_k^{\alpha_k},$$

where $p_1, p_2, \ldots, p_k$ are distinct prime numbers and $\alpha_1, \alpha_2, \ldots, \alpha_k$ are positive integers. Then the number of positive factors of n is

$$\tau(n) = (\alpha_1 + 1)(\alpha_2 + 1) \cdots (\alpha_k + 1).$$

(4) *The formula for calculating the sum of positive factors*

Suppose the prime factorization of n is

$$n = p_1^{\alpha_1} p_2^{\alpha_2} \cdots p_k^{\alpha_k},$$

where $p_1, p_2, \ldots, p_k$ are distinct prime numbers and $\alpha_1, \alpha_2, \ldots, \alpha_k$ are positive integers. Then the sum of positive factors of n is

$$\sigma(n) = \prod_{i=1}^{k} (1 + p_i + \cdots + p_i^{\alpha_i}) = \prod_{i=1}^{k} \frac{p_i^{\alpha_i+1} - 1}{p_i - 1}.$$

Example 1.2. Suppose some positive integer n has $\tau(n) = 6$ positive factors. Find all possible values of $\tau(n^2)$, which denotes the number of positive factors of n^2. Determine the least possible value of $\sigma(n)$, which denotes the sum of positive factors of n.

Solution. Evidently $n > 1$. Suppose the prime factorization of n is $n = p_1^{\alpha_1} \cdots p_k^{\alpha_k}$, where $p_1 < \cdots < p_k$ are distinct prime numbers. Then by the given condition,

$$\tau(n) = \prod_{i=1}^{k} (1 + \alpha_i) = 6.$$

Observe that $1 + \alpha_i \geq 2 (1 \leq i \leq k)$, so $k \leq 2$.
    When $k = 1$, we have $a_1 = 5$ and $n = p_1^5$. Hence

$$\tau(n^2) = \tau(p_1^{10}) = 11,$$
$$\sigma(n) = 1 + p_1 + \cdots p_1^5 \geq 1 + 2 + \cdots + 2^5 = 63.$$

When $k = 2$, we see that α_1, α_2 is a permutation of $1, 2$, which implies that $n = p_1^2 p_2$ or $p_1 p_2^2$, where $p_1 < p_2$.
    If $n = p_1^2 p_2$, then

$$\tau(n^2) = \tau(p_1^4 p_2^2) = (1+4)(1+2) = 15,$$
$$\sigma(n) = (1 + p_1 + p_1^2)(1 + p_2) \geq (1 + 2 + 2^2)(1 + 3) = 28.$$

The equality holds when $p_1 = 2, p_2 = 3$ (i.e., $n = 12$).
    If $n = p_1 p_2^2$, then similarly $\tau(n^2) = 15$ and $\sigma(n) \geq (1+2)(1+3+3^2) = 39$.
    Summarizing, all possible values of $\tau(n^2)$ are 11 and 15, and the least possible value of $\sigma(n)$ is 28.
    Sometimes we want to study the power exponent of some prime p in the prime factorization of a positive integer n. In this book, it will be denoted as $v_{p_i}(n) = \alpha_i$ (specifically if n is not divisible by p, then $v_p(n) = 0$). When handling problems about divisibility of integers, an important conclusion is: The relation $a \mid b$ is equivalent to that $v_p(a) \leq v_p(b)$ for all prime numbers p.

(5) *The formula for factorial and binomial coefficients*

For a positive integer n, the factorial of n, denoted as $n!$, is the product of $1, 2, \ldots, n$. We also define $0! = 1$.

Sometimes we study the power exponent of some prime p in the prime factorization of a positive integer n. The formula is

$$v_p(n!) = \sum_{i=1}^{\infty} \left[\frac{n}{p^i} \right],$$

where $[x]$ represents the greatest integer less than or equal to x.

In earlier IMOs, there were relatively many problems discussing factorial and binomial coefficients. A binomial coefficient has the form

$$C_n^m = \frac{n!}{m!(n-m)!},$$

where m, n are positive integers and $m \le n$.

(6) *The formula for positional notations*

A positional system is a number system in which the contribution of a digit to the value of a number is the value of the digit multiplied by a factor determined by the position of the digit. The most commonly used positional system is the number system with base ten. We denote an n-digit number in the system with base r by $\overline{a_1 a_2 \cdots a_n}_{(r)}$, whose value in the system with base ten is

$$a_1 \cdot r^{n-1} + a_2 \cdot r^{n-2} + \cdots + a_n \cdot r^0.$$

In earlier IMOs, there were several problems discussing positional notations. Some other problems related to divisibility of integers have concise solutions by considering positional notations.

Example 1.3. For a given real number x, find the value of $S(x) = \sum_{n=1}^{+\infty} \frac{(-1)^{[2^n x]}}{2^n}$.

Solution. Suppose the base 2 (or binary) representation of x is

$$x = (\overline{a_k a_{k-1} \ldots a_0 . b_1 b_2 \ldots})_2.$$

Then $a_k, a_{k-1}, \ldots, a_0, b_1, b_2, \ldots$ are all 0 or 1. If x is an integer or a terminating decimal, then starting from some term, $b_i, b_{i+1}, b_{i+2}, \ldots$ are all zeros.

For any positive integer n, note that

$$[2^n x] = (\overline{a_k a_{k-1} \cdots a_0 b_1 b_2 \cdots b_n})_2 \equiv b_n \pmod 2,$$

and so $(-1)^{[2^n x]} = (-1)^{b_n}$. No matter $b_n = 0, 1$, we always have $(-1)^{b_n} = 1 - 2b_n$. Thus

$$(-1)^{[2^n x]} = (-1)^{b_n} = 1 - 2b_n.$$

Therefore,

$$S(x) = \sum_{n=1}^{\infty} \frac{1 - 2b_n}{2^n} = \sum_{n=1}^{\infty} \frac{1}{2^n} - 2 \times \sum_{n=1}^{\infty} \frac{b_n}{2^n} = 1 - 2 \times (\overline{0.b_1 b_2 \cdots})_2$$

$$= 1 - 2(x - [x]) = 1 + 2[x] - 2x.$$

(7) *The Gauss function*

For a real number x, the integral part of x is the greatest integer less than or equal to x, denoted as $[x]$. The function $y = [x]$ is called the floor function. The square bracket notation is introduced by Gauss, and hence we also call it the Gauss function. Furthermore, we denote $x - [x]$ by $\{x\}$, i.e. $\{x\} = x - [x]$, which is called the fractional part of x.

Property 1.1. For any real number x and integer n,

$$[n + x] = n + [x].$$

Property 1.2. For any real numbers x and y,

$$[x] + [y] \le [x + y], \quad \text{and} \quad \{x\} + \{y\} \ge \{x + y\}.$$

Moreover, for real numbers $x_i (i = 1, 2, \ldots, n)$,

$$[x_1 + x_2 + \cdots + x_n] \ge [x_1] + [x_2] + \cdots + [x_n].$$

In particular, when $x_1 = x_2 = \cdots = x_n = x$,

$$[nx] \ge n[x] \ (x \text{ is a real number and } n \text{ is a positive integer}).$$

Property 1.3. For any real number x and positive integer n,

$$\left[\frac{x}{n}\right] = \left[\frac{[x]}{n}\right].$$

Property 1.4. In the first n consecutive integers $1, 2, \ldots, n$, there are $\left[\frac{n}{p}\right]$ numbers that are divisible by a positive integer p.

Property 1.5. In the prime factorization of $n!$, the power of prime p is

$$v_p(n!) = \left[\frac{n}{p}\right] + \left[\frac{n}{p^2}\right] + \cdots + \left[\frac{n}{p^m}\right],$$

where $p^m \leq n$, and $p^{m+1} > n$.

Example 1.4. Prove that for any real number x, we have $[x] + \left[x + \frac{1}{2}\right] = [2x]$.

Proof.   Suppose $x = [x] + \{x\}$, where $0 \leq \{x\} < 1$.
   Then $x + \frac{1}{2} = [x] + \{x\} + \frac{1}{2}, \quad 2x = 2[x] + 2\{x\}$.
   When $0 \leq \{x\} < \frac{1}{2}$,

$$\frac{1}{2} \leq \{x\} + \frac{1}{2} < 1, \quad 0 \leq 2\{x\} < 2 \times \frac{1}{2} = 1,$$

and then

$$\left[x + \frac{1}{2}\right] = [x], \quad [2x] = 2[x].$$

Hence,

$$[x] + \left[x + \frac{1}{2}\right] = 2[x] = [2x].$$

When $\frac{1}{2} \leq \{x\} < 1$,

$$1 \leq \{x\} + \frac{1}{2} < \frac{3}{2}, \quad 1 \leq 2\{x\} < 2 \times 1 = 2,$$

and then

$$\left[x + \frac{1}{2}\right] = [x] + \left[\{x\} + \frac{1}{2}\right] = [x] + 1, \quad [2x] = 2[x] + [2\{x\}] = 2[x] + 1.$$

   Hence,

$$[x] + \left[x + \frac{1}{2}\right] = 2[x] + 1 = [2x].$$

As a conclusion, $[x] + \left[x + \frac{1}{2}\right] = [2x]$ for any real number x.

Remark. This equation has a more general form: for any real number x and any integer greater than 1,

$$[x] + \left[x + \frac{1}{n}\right] + \left[x + \frac{2}{n}\right] + \cdots + \left[x + \frac{n-1}{n}\right] = [nx].$$

This equality is called Hermite's identity.

1.2 Problems and Solutions

1.2.1 *Discussion on divisiblility*

Problem 1.1 (IMO 1-1, proposed by Poland). Prove that the fraction $\frac{21n+4}{14n+3}$ is irreducible for every natural number n.

Proof 1. Let d be the greatest common factor of $21n + 4$ and $14n + 3$. Then

$$d = \gcd(21n + 4, 14n + 3) = \gcd(7n + 1, 14n + 3) = \gcd(7n + 1, 1) = 1.$$

Hence the fraction $\frac{21n+4}{14n+3}$ is irreducible.

Proof 2. Since $3(14n + 3) - 2(21n + 4) = 1$, the numerator is relatively prime to the denominator. Then the fraction $\frac{21n+4}{14n+3}$ is irreducible.

Remark. As the first problem of the first IMO, it does not look difficult from today's perspective. However, the discussion on properties of common divisors and coprime numbers was the foundation of several difficult number theory problems in later IMOs.

【Score Situation】 This particular problem saw the following distribution of scores among contestants: 4 contestants scored 5 points, no contestant scored 4 points, no contestant scored 3 points, 1 contestant scored 2 points, 3 contestants scored 1 point, and 1 contestant scored 0 point. The average score of this problem is 2.778, indicating that it had a certain level of difficulty.

Among the top five teams in the team scores, the Romania team achieved a total score of 249 points, the Hungary team achieved a total score of 233 points, the Czechoslovakia team achieved a total score of 192 points, the Bulgaria team achieved a total score of 131 points, and the Poland team achieved a total score of 122 points.

The gold medal cutoff for this IMO was set at 37 points (with 3 contestants earning gold medals), the silver medal cutoff was 36 points (with 3 contestants earning silver medals), and the bronze medal cutoff was 33 points (with 5 contestants earning bronze medals).

In this IMO, only one contestant achieved a perfect score of 40 points, namely Bohuslav Diviš from Czechoslovakia.

Problem 1.2 (IMO 9-3, proposed by the United Kingdom). Let k, m, and n be natural numbers such that $m + k + 1$ is a prime greater than $n + 1$. Let $C_s = s(s+1)$. Prove that the product

$$(C_{m+1} - C_k)(C_{m+2} - C_k) \cdots (C_{m+n} - C_k)$$

is divisible by the product $C_1 C_2 \cdots C_n$.

Proof.   For natural numbers s and t,

$$C_s - C_t = s^2 + s - t^2 - t = (s - t)(s + t + 1).$$

Hence, the product

$$(C_{m+1} - C_k)(C_{m+2} - C_k) \cdots (C_{m+n} - C_k)$$
$$= (m - k + 1)(m - k + 2) \cdots (m - k + n)$$
$$\cdot (m + k + 2)(m + k + 3) \cdots (m + k + n + 1).$$

Also we can easily find that the product $C_1 C_2 \cdots C_n = n! \cdot (n+1)!$.

Note that the product of any n consecutive integers is divisible by $n!$. Then

$$n! \mid (m - k + 1)(m - k + 2) \cdots (m - k + n),$$
$$(n+1)! \mid (m + k + 1)(m + k + 2) \cdots (m + k + n + 1).$$

We are told that $m + k + 1$ is a prime greater than $n + 1$, so it is relatively prime to $(n+1)!$. Then $(m+k+2)(m+k+3) \cdots (m+k+n+1)$ is divisible by $(n + 1)!$. As a result,

$$(m - k + 1)(m - k + 2) \cdots (m - k + n) \cdot$$
$$(m + k + 2)(m + k + 3) \cdots (m + k + n + 1)$$

is divisible by $n! \cdot (n+1)!$.

Remark. In this proof, we used the conclusion that the product of n consecutive integers is divisible by $n!$, which is quite useful when dealing with binomial coefficients. Interestingly, this conclusion itself can be proven by the definition of binomial coefficients. Of course, it is also feasible to prove this problem by discussing the power exponent of each prime in the factorials. This skill will be shown in future problems.

【Score Situation】 This particular problem saw the following distribution of scores among contestants: 12 contestants scored 8 points, no contestant scored 7 points, 1 contestant scored 6 points, no contestant scored 5 points, 2 contestants scored 4 points, 2 contestants scored 3 points, 3 contestant scored 2 points, 1 contestant scored 1 point, and 16 contestants scored 0 point. The average score of this problem is 3.324, indicating that it was relatively straightforward.

Among the top five teams in the team scores, the Soviet Union team achieved a total score of 275 points, the German Democratic Republic team achieved a total score of 257 points, the Hungary team achieved a total score of 251 points, the United Kingdom team achieved a total score of 231 points, and the Romania team achieved a total score of 214 points.

The gold medal cutoff for this IMO was set at 38 points (with 11 contestants earning gold medals), the silver medal cutoff was 30 points (with 14 contestants earning silver medals), and the bronze medal cutoff was 22 points (with 26 contestants earning bronze medals).

In this IMO, a total of five contestants achieved a perfect score of 42 points.

Problem 1.3 (IMO 14-3, proposed by the United Kingdom). Let m and n be arbitrary non-negative integers. Prove that

$$\frac{(2m)!(2n)!}{m!n!(m+n)!}$$

is an integer ($0! = 1$).

Proof 1. We first prove a lemma.

Lemma. *For any non-negative real x and y, the inequality $[x] + [y] + [x+y] \le [2x] + [2y]$ holds.*

Proof of the lemma. Let $\{x\} = x - [x]$ and $\{y\} = y - [y]$. Then

$$[x] + [y] + [x+y] = 2[x] + 2[y] + [\{x\} + \{y\}],$$
$$[2x] + [2y] = 2[x] + 2[y] + [2\{x\}] + [2\{y\}],$$

so it suffices to prove the inequality for $0 \le x, y < 1$.

When $0 \le x+y < 1$, we have $[x] + [y] + [x+y] = 0 \le [2x] + [2y]$. When $1 \le x + y < 2$, at least one of x and y is not less than $\frac{1}{2}$, which means $[2x] + [2y] \ge 1$, and then $[x] + [y] + [x+y] = 1 \le [2x] + [2y]$. The lemma is proved.

Back to the original problem, to show that $\frac{(2m)!(2n)!}{m!n!(m+n)!}$ is an integer, we show that for any prime p,

$$v_p((2m)!) + v_p((2n)!) \geq v_p(m!) + v_p(n!) + v_p((m+n)!). \qquad (*)$$

In this inequality,

$$v_p\left((2m)!\right) + v_p\left((2n)!\right) = \sum_{i=1}^{\infty}\left(\left[\frac{2m}{p^i}\right] + \left[\frac{2n}{p^i}\right]\right),$$

$$v_p\left(m!\right) + v_p\left(n!\right) + v_p\left((m+n)!\right) = \sum_{i=1}^{\infty}\left(\left[\frac{m}{p^i}\right] + \left[\frac{n}{p^i}\right] + \left[\frac{m+n}{p^i}\right]\right).$$

By the lemma, $\left[\frac{2m}{p^i}\right] + \left[\frac{2n}{p^i}\right] \geq \left[\frac{m}{p^i}\right] + \left[\frac{n}{p^i}\right] + \left[\frac{m+n}{p^i}\right]$ for any $i \in \mathbf{N}^*$. Then

$$\sum_{i=1}^{\infty}\left(\left[\frac{2m}{p^i}\right] + \left[\frac{2n}{p^i}\right]\right) \geq \sum_{i=1}^{\infty}\left(\left[\frac{m}{p^i}\right] + \left[\frac{n}{p^i}\right] + \left[\frac{m+n}{p^i}\right]\right).$$

Hence, $(*)$ is proved and therefore $\frac{(2m)!(2n)!}{m!n!(m+n)!}$ is an integer.

Proof 2. To prove the given expression is an integer, let $f(m,n) = \frac{(2m)!(2n)!}{m!n!(m+n)!}$. For any non-negative integer m, $f(m,0) = \frac{(2m)!}{m!m!} = C_{2m}^{m}$ is an integer. For some non-negative integer n, if $f(m,n)$ is an integer for any non-negative integer m, then

$$f\left(m, n+1\right) = \frac{(2m)!\,(2n+2)!}{m!\,(n+1)!\,(m+n+1)!} = \frac{(2m)!\,(2n)!}{m!n!\,(m+n)!} \cdot \frac{4n+2}{m+n+1}$$

$$= \frac{(2m)!\,(2n)!}{m!n!\,(m+n)!} \cdot \left[4 - \frac{(2m+2)\,(2m+1)}{(m+1)\,(m+n+1)}\right]$$

$$= 4 \cdot \frac{(2m)!\,(2n)!}{m!n!\,(m+n)!} - \frac{(2m+2)!\,(2n)!}{(m+1)!n!\,(m+1+n)!}$$

$$= 4f(m,n) - f(m+1,n),$$

which shows that $f(m, n+1)$ is an integer for any non-negative m. By mathematical induction, the given expression $f(m,n)$ is an integer for any non-negative m,n.

Remark. We provided a straightforward proof first. When dealing with factorials and divisibility problems, the first thing we can try is to study the power of prime factors. On the other hand, the product and quotient of factorials lead us to think of the binomial coefficients. Then similarly to the recurrence relation of binomial coefficients, we can get the recurrence formula in the second proof.

【Score Situation】 This particular problem saw the following distribution of scores among contestants: 10 contestants scored 7 points, no contestant scored 6 points, no contestant scored 5 points, no contestant scored 4 points, 1 contestant scored 3 points, 1 contestant scored 2 points, 2 contestants scored 1 point, and 19 contestants scored 0 point. The average score of this problem is 2.333, indicating that it had a certain level of difficulty.

Among the top five teams in the team scores, the scores of this problem are as follows: The Soviet Union team scored 35 points (with a total team score of 270 points), the Hungary team scored 43 points (with a total team score of 263 points), the German Democratic Republic team scored 45 points (with a total team score of 239 points), the Romania team scored 47 points (with a total team score of 208 points), and the United Kingdom team scored 21 points (with a total team score of 179 points).

The gold medal cutoff for this IMO was set at 40 points (with 8 contestants earning gold medals), the silver medal cutoff was 30 points (with 16 contestants earning silver medals), and the bronze medal cutoff was 19 points (with 30 contestants earning bronze medals).

In this IMO, a total of eight contestants achieved a perfect score of 40 points.

Problem 1.4 (IMO 19-3, proposed by the Netherlands). Let n be a given integer >2, and let V_n be the set of integers $1+kn$, where $k = 1, 2, \ldots$. A number $m \in V_n$ is called *indecomposable* in V_n if there do not exist numbers $p, q \in V_n$ such that $pq = m$. Prove that there exists a number $r \in V_n$ that can be expressed as the product of elements indecomposable in V_n in more than one way (products which differ only in the order of their factors will be considered the same).

Proof. Let $a = n-1$, $b = 2n-1$, and $r = a^2 b^2$. Then $a^2, b^2, ab, r \in V_n$, for $a^2 \equiv b^2 \equiv ab \equiv a^2 b^2 \equiv 1 \pmod{n}$. On the other hand, a^2 is indecomposable in V_n because $a^2 < (n+1)^2$.

Furthermore, $\frac{ab}{a^2} = \frac{b}{a} = \frac{2n-1}{n-1} = 2 + \frac{1}{n-1}$ is not an integer for $n > 2$, so a^2 is not a divisor of ab.

Now we have at least two distinct factorings of r into indecomposables: one contains a^2, using $r = a^2 \cdot b^2$ and then factoring b^2 into indecomposables; the other one does not contain a^2, using $r = (ab)^2$ and then factoring ab into indecomposables.

Remark. A considerable number of high school math problems have a background in higher mathematics, especially those number theory problems. This problem from the 19th IMO is one example. For high school students without the knowledge of abstract algebra (including groups and rings), the key to this problem is to check some simple cases and try to find such forms $a^2 \cdot b^2 = ab \cdot ab$ so that we can factor a number in two ways.

【Score Situation】 This particular problem saw the following distribution of scores among contestants: 11 contestants scored 7 points, 8 contestants scored 6 points, no contestant scored 5 points, no contestant scored 4 points, 1 contestant scored 3 points, no contestant scored 2 points, no contestant scored 1 point, and 17 contestants scored 0 point. The average score of this problem is 3.459, indicating that it was relatively straightforward.

Among the top five teams in the team scores, the United States team achieved a total score of 202 points, the Soviet Union team achieved a total score of 192 points, the Hungary team achieved a total score of 190 points, the United Kingdom team achieved a total score of 190 points, and the Netherlands team achieved a total score of 185 points.

The gold medal cutoff for this IMO was set at 34 points (with 13 contestants earning gold medals), the silver medal cutoff was 24 points (with 29 contestants earning silver medals), and the bronze medal cutoff was 17 points (with 35 contestants earning bronze medals).

In this IMO, a total of five contestants achieved a perfect score of 40 points.

Problem 1.5 (IMO 21-1, proposed by Germany). Let p and q be natural numbers such that

$$\frac{p}{q} = 1 - \frac{1}{2} + \frac{1}{3} - \cdots - \frac{1}{1318} + \frac{1}{1319}.$$

Prove that p is divisible by 1979.

Proof.

$$\frac{p}{q} = 1 - \frac{1}{2} + \frac{1}{3} - \cdots - \frac{1}{1318} + \frac{1}{1319}$$

$$= 1 + \frac{1}{2} + \frac{1}{3} + \cdots + \frac{1}{1318} + \frac{1}{1319} - 2\left(\frac{1}{2} + \frac{1}{4} + \cdots + \frac{1}{1318}\right)$$

$$= 1 + \frac{1}{2} + \frac{1}{3} + \cdots + \frac{1}{1318} + \frac{1}{1319} - \left(1 + \frac{1}{2} + \cdots + \frac{1}{659}\right)$$

$$= \frac{1}{660} + \frac{1}{661} + \cdots + \frac{1}{1318} + \frac{1}{1319}$$

$$= \left(\frac{1}{660} + \frac{1}{1319} \right) + \left(\frac{1}{661} + \frac{1}{1318} \right) \cdots + \left(\frac{1}{989} + \frac{1}{990} \right)$$

$$= 1979 \left(\frac{1}{660 \times 1319} + \frac{1}{661 \times 1318} + \frac{1}{989 \times 990} \right)$$

$$= \frac{1979N}{1319!}.$$

Now, $1979 \mid p \cdot 1319!$, and since 1979 is a prime (we can get this from checking the primes not larger than 43), we have $(1979, 1319!) = 1$. Then $1979 \mid p$.

Remark. Here, the most important part is the identity $\sum_{k=1}^{2n} (-1)^{k-1} \cdot \frac{1}{k} = \sum_{k=n+1}^{2n} \frac{1}{k}$.

【Score Situation】 This particular problem saw the following distribution of scores among contestants: 39 contestants scored 6 points, 1 contestant scored 5 points, 1 contestant scored 4 points, 2 contestants scored 3 points, 7 contestants scored 2 points, 11 contestants scored 1 point, and 105 contestants scored 0 point. The average score of this problem is 1.651, indicating that it was relatively challenging.

Among the top five teams in the team scores, the scores of this problem are as follows: The Soviet Union team scored 36 points (with a total team score of 267 points), the Romania team scored 31 points (with a total team score of 240 points), the Germany team scored 36 points (with a total team score of 235 points), the United Kingdom team scored 10 points (with a total team score of 218 points), and the United States team scored 9 points (with a total team score of 199 points).

The gold medal cutoff for this IMO was set at 37 points (with 8 contestants earning gold medals), the silver medal cutoff was 29 points (with 32 contestants earning silver medals), and the bronze medal cutoff was 20 points (with 42 contestants earning bronze medals).

In this IMO, a total of four contestants achieved a perfect score of 40 points.

Problem 1.6 (IMO 22-4, proposed by Belgium). (a) For which values of $n > 2$ is there a set of n consecutive positive integers such that the largest number in the set is a divisor of the least common multiple of the remaining $n - 1$ numbers?

(b) For which values of $n > 2$ is there exactly one set having the stated property?

Solution. Let the n consecutive positive integers be $m - n + 1, m - n + 2, \ldots, m$.

For $n = 3$, the condition $m \mid (m - 1)(m - 2)$ implies $m \mid 2$, which is in contradiction to $m \geq n = 3$.

For $n = 4$, from $m \mid (m-1)(m-2)(m-3)$, one has $m \mid 6$. Since $m \geq n = 4$, it follows that $m = 6$. We have exactly one set of 4 consecutive positive integers: 3,4,5,6.

Suppose $n > 4$. For $(n-1)(n-2) > n - 1$, we can let $m = (n-1)(n-2)$. Now $(n-1) \mid m - (n-1)$ and $(n-2) \mid m - (n-2)$. Since $(m - (n-1), m - (n-2)) = 1$, we know that $m = (n-1)(n-2)$ divides the least common multiple of $m - n + 1$ and $m - n + 2$.

We can also let $m = (n-2)(n-3)$ because $(n-2)(n-3) > n - 1$. Similarly, we have that $m = (n-2)(n-3)$ divides the least common multiple of $m - n + 2$ and $m - n + 3$.

Thus for $n > 4$ we can find more than one admissible sequence of n consecutive positive integers. The answers to (a) and (b) are $n \geq 4$ and $n = 4$, respectively.

Remark. For many problems, we can study some simple cases first. In this solution, such strategy is beneficial to the contradiction for $n = 3$, the proof of unique existence for $n = 4$, and the construction for $n > 4$.

【Score Situation】 This particular problem saw the following distribution of scores among contestants: 37 contestants scored 7 points, 4 contestants scored 6 points, 4 contestants scored 5 points, 3 contestants scored 4 points, 1 contestant scored 3 points, 1 contestant scored 2 points, no contestant scored 1 point, and 1 contestant scored 0 point. The average score of this problem is 6.275, indicating that it was simple.

Among the top five teams in the team scores, the scores of this problem are as follows: The United States team scored 56 points (with a total team score of 314 points), the Germany team scored 54 points (with a total team score of 312 points), the United Kingdom team scored 54 points (with a total team score of 301 points), the Austria team scored 51 points (with a total team score of 290 points), and the Bulgaria team scored 50 points (with a total team score of 287 points).

The gold medal cutoff for this IMO was set at 41 points (with 36 contestants earning gold medals), the silver medal cutoff was 34 points (with 37 contestants earning silver medals), and the bronze medal cutoff was 26 points (with 30 contestants earning bronze medals).

In this IMO, a total of 26 contestants achieved a perfect score of 42 points.

Problem 1.7 (IMO 25-2, proposed by the Netherlands). Find one pair of positive integers a and b such that:

(i) $ab(a + b)$ is not divisible by 7;

(ii) $(a + b)^7 - a^7 - b^7$ is divisible by 7^7.

Justify your answer.

Solution. Since

$$(a+b)^7 - a^7 - b^7$$

$$= 7a^6b + 21a^5b^2 + 35a^4b^3 + 35a^3b^4 + 21a^2b^5 + 7ab^6$$

$$= 7ab\,(a+b)\,\left(a^4 + 2a^3b + 3a^2b^2 + 2ab^3 + b^4\right)$$

$$= 7ab\,(a+b)\,\left(a^2 + ab + b^2\right)^2,$$

and $ab(a + b)$ is not divisible by 7, then (ii) holds if and only if $7^3 \mid a^2 + ab + b^2$. Now that $7^3 = 343 = 18 \times 19 + 1 = 18^2 + 18 + 1$, so we can let $a = 18$ and $b = 1$, which also satisfies (i). Summarizing, $(a, b) = (18, 1)$ is a solution.

【Score Situation】 This particular problem saw the following distribution of scores among contestants: 66 contestants scored 7 points, 3 contestants scored 6 points, 6 contestants scored 5 points, 7 contestants scored 4 points, 8 contestants scored 3 points, 8 contestants scored 2 points, 39 contestants scored 1 point, and 55 contestants scored 0 point. The average score of this problem is 3.214, indicating that it was relatively straightforward.

Among the top five teams in the team scores, the scores of this problem are as follows: The Soviet Union team scored 35 points (with a total team score of 235 points), the Bulgaria team scored 34 points (with a total team score of 203 points), the Romania team scored 33 points (with a total team score of 199 points), the Hungary team scored 42 points (with a total team score of 195 points), and the United States team scored 37 points (with a total team score of 195 points).

The gold medal cutoff for this IMO was set at 40 points (with 14 contestants earning gold medals), the silver medal cutoff was 26 points (with 35 contestants earning silver medals), and the bronze medal cutoff was 17 points (with 49 contestants earning bronze medals).

In this IMO, a total of eight contestants achieved a perfect score of 42 points.

Problem 1.8 (IMO 25-6, proposed by Poland). Let a, b, c, and d be odd integers such that $0 < a < b < c < d$ and $ad = bc$. Prove that if $a + d = 2^k$ and $b + c = 2^m$ for some integers k and m, then $a = 1$.

Proof. Since $ad = bc$,

$$a\,((a+d) - (b+c)) = a^2 + ad - ab - ac$$

$$= a^2 + bc - ab - ac$$

$$= (a - b)(a - c) > 0,$$

so $a + d > b + c$. Then $2^k > 2^m$ with $k > m$.

Since $a(2^k - a) = ad = bc = b(2^m - b)$, we have

$$2^m \left(b - 2^{k-m}a\right) = b^2 - a^2 = (b+a)(b-a),$$

and then $2^m \mid (b+a)(b-a)$. We observe that $(b+a) - (b-a) = 2a$, and a is odd, which means $2a$ is not divisible by 4, so at least one of $b+a$ and $b-a$ is not divisible by 4. Hence $2^{m-1} \mid b+a$ or $2^{m-1} \mid b-a$. Since $0 < b-a < b < \frac{1}{2}(b+c) = 2^{m-1}$, then $2^{m-1} \mid b+a$. However, $0 < b+a < b+c = 2^m$, so $b+a = 2^{m-1}$, from which $b = 2^{m-1} - a$ and $c = 2^{m-1} + a$.

Now $ad = bc = (2^{m-1} - a)(2^{m-1} + a) = 2^m - a^2$, which leads to $a(a+d) = 2^m$. Therefore, $a \mid 2^m$. For a is odd, $a = 1$.

Remark. When it comes to the power of 2, or the power of other prime numbers, such as p^α, a common approach is to rearrange all the prime factors to one side of the equality and consider whether we have $p^\alpha \mid n$ or $n \mid p^\alpha$. For instance, in solving problem 1.8 we used $2^m(b - 2^{k-m}a) = (b+a)(b-a)$ and $a(a+d) = 2^m$ to obtain $2^m \mid (b+a)(b-a)$ and $a \mid 2^m$, and then we got the desired conclusion.

【Score Situation】 This particular problem saw the following distribution of scores among contestants: 25 contestants scored 7 points, 3 contestants scored 6 points, 3 contestants scored 5 points, 5 contestants scored 4 points, 6 contestants scored 3 points, 9 contestants scored 2 points, 18 contestants scored 1 point, and 123 contestants scored 0 point. The average score of this problem is 1.469, indicating that it was relatively challenging.

Among the top five teams in the team scores, the scores of this problem are as follows: The Soviet Union team scored 42 points (with a total team score of 235 points), the Bulgaria team scored 30 points (with a total team score of 203 points), the Romania team scored 21 points (with a total team score of 199 points), the United States team scored 21 points (with a total team score of 195 points), and the Hungary team scored 15 points (with a total team score of 195 points).

The gold medal cutoff for this IMO was set at 40 points (with 14 contestants earning gold medals), the silver medal cutoff was 26 points (with 35 contestants earning silver medals), and the bronze medal cutoff was 17 points (with 49 contestants earning bronze medals).

In this IMO, a total of eight contestants achieved a perfect score of 42 points.

Problem 1.9 (IMO 33-1, proposed by New Zealand). Find all integers a, b, and c with $1 < a < b < c$ such that $(a-1)(b-1)(c-1)$ is a divisor of $abc - 1$.

Solution. Let $x = a - 1$, $y = b - 1$, and $z = c - 1$. Then $x < y < z$ are positive integers and their product xyz is a divisor of

$$(x + 1)(y + 1)(z + 1) - 1 = xyz + xy + yz + zx + x + y + z.$$

We have $xyz \mid (xy + yz + zx + x + y + z)$. If $x \geq 3$, then $y \geq 4$ and $z \geq 5$. Hence

$$1 \leq \frac{xy + yz + zx + x + y + z}{xyz}$$

$$= \frac{1}{x} + \frac{1}{y} + \frac{1}{z} + \frac{1}{xy} + \frac{1}{yz} + \frac{1}{zx}$$

$$\leq \frac{1}{3} + \frac{1}{4} + \frac{1}{5} + \frac{1}{12} + \frac{1}{20} + \frac{1}{15} = \frac{59}{60} < 1,$$

which is a contradiction. So, $x \leq 2$.

If $x = 1$, then $yz \mid y + yz + z + 1 + y + z$, that is $yz \mid 2y + 2z + 1$. So y and z are both odd. Now that $2z \leq yz \leq 2y + 2z + 1 < 4z$, so $2 \leq y < 4$. Since y is odd, $y = 3$. We also have $3z \mid 2z + 7$, from which $z \mid 7$, and then $z = 7$.

If $x = 2$, then $2yz \mid 2y + yz + 2z + 2 + y + z$, namely $2yz \mid yz + 3y + 3z + 2$. So y and z are both even. Hence, $y \geq 4$ and $z \geq y + 2$. Now that

$$2yz \leq yz + 3y + 3z + 2$$

$$\leq yz + 3(z - 2) + 3z + 2$$

$$< yz + 6z,$$

so $2 = x < y < 6$. The fact that y is even implies $y = 4$. We have $8z \mid 7z + 14$, in other words $z \mid 14$. Then since z is even and $z > y = 4$, we know $z = 14$.

Now, we can see that $(x, y, z) = (1, 3, 7), (2, 4, 14)$ both satisfy the given condition, so the solution is $(a, b, c) = (2, 4, 8), (3, 5, 15)$.

Remark. For problems concerning divisions of integers, we usually want the divisors in a simpler form such as x, y or p, q. This is why we let $x = a - 1$, $y = b - 1$, and $z = c - 1$ at first. Moreover, since the coefficient of the highest order term in the expression of the dividend is less than that in the divisor, we know that for larger x, y, z, the dividend will be less than the divisor. Therefore we only had to discuss those smaller cases.

【Score Situation】 This particular problem saw the following distribution of scores among contestants: 131 contestants scored 7 points, 18 contestants scored 6 points, 20 contestants scored 5 points, 14 contestants scored 4 points, 14 contestants scored 3 points, 11 contestants scored 2 points, 32 contestants scored 1 point, and 110 contestants scored 0 point. The average score of this problem is 3.649, indicating that it was relatively straightforward.

Among the top five teams in the team scores, the scores of this problem are as follows: The China team scored 41 points (with a total team score of 240 points), the United States team scored 37 points (with a total team score of 181 points), the Romania team scored 38 points (with a total team score of 177 points), the Commonwealth of Independent States team scored 41 points (with a total team score of 176 points), and the United Kingdom team scored 42 points (with a total team score of 168 points).

The gold medal cutoff for this IMO was set at 32 points (with 26 contestants earning gold medals), the silver medal cutoff was 24 points (with 55 contestants earning silver medals), and the bronze medal cutoff was 14 points (with 74 contestants earning bronze medals).

In this IMO, a total of four contestants achieved a perfect score of 42 points.

Problem 1.10 (IMO 35-4, proposed by Australia). Determine all ordered pairs (m, n) of positive integers such that

$$\frac{n^3 + 1}{mn - 1}$$

is an integer.

Solution. Since $mn - 1 \mid (mn)^3 - 1$ for any positive integers m, n, and m, n satisfy $mn - 1 \mid n^3 + 1$, we have $mn - 1 \mid (mn)^3 + n^3 = n^3(m^3 + 1)$.

We observe that $(mn - 1, n) = 1$, so $mn - 1 \mid m^3 + 1$, which implies that $mn - 1 \mid n^3 + 1$ is equivalent to $mn - 1 \mid m^3 + 1$. Thus by symmetry we can assume $m \geq n$.

When $n = 1$, the condition becomes $mn - 1 = m - 1$ is a divisor of $1^3 + 1 = 2$, so $m = 2$ or 3.

When $n \geq 2$, if $m = n$, then the condition becomes $n^2 - 1 \mid n^3 + 1$, that is $n - 1 \mid n^2 - n + 1$, from which $n - 1 \mid 1$, so $n = 2$ and $m = 2$.

If $m > n$, then $m \geq n + 1$. Now from the condition, $mn - 1 \mid n^3 + mn - (mn - 1)$. Then

$$mn - 1 \mid n^3 + mn = n(n^2 + m).$$

Using $(mn - 1, n) = 1$, we conclude that $mn - 1 \mid n^2 + m$.

On the other hand,

$$2(mn - 1) \geq n(n + 1) + mn - 2 = n^2 + mn + n - 2 > n^2 + m,$$

which indicates that $mn - 1 = n^2 + m$. Hence $(m - n - 1)(n - 1) = 2$.
Now we have that

$$\begin{cases} m - n - 1 = 1, \\ n - 1 = 2, \end{cases} \quad \text{or} \quad \begin{cases} m - n - 1 = 2, \\ n - 1 = 1. \end{cases}$$

The solution is $m = 5$, $n = 3$ or $m = 5$, $n = 2$.
Consequently,

$$(m, n) = (2, 1), (3, 1), (1, 2), (1, 3), (2, 2), (5, 2), (5, 3), (2, 5), (3, 5).$$

We can see these pairs (m, n) all satisfy the given condition.

【Score Situation】 This particular problem saw the following distribution of scores among contestants: 101 contestants scored 7 points, 11 contestants scored 6 points, 20 contestants scored 5 points, 19 contestants scored 4 points, 43 contestants scored 3 points, 70 contestants scored 2 points, 69 contestants scored 1 point, and 52 contestants scored 0 point. The average score of this problem is 3.343, indicating that it was relatively straightforward.

Among the top five teams in the team scores, the scores of this problem are as follows: The United States team scored 42 points (with a total team score of 252 points), the China team scored 42 points (with a total team score of 229 points), the Russia team scored 42 points (with a total team score of 224 points), the Bulgaria team scored 34 points (with a total team score of 223 points), and the Hungary team scored 37 points (with a total team score of 221 points).

The gold medal cutoff for this IMO was set at 40 points (with 30 contestants earning gold medals), the silver medal cutoff was 30 points (with 64 contestants earning silver medals), and the bronze medal cutoff was 19 points (with 98 contestants earning bronze medals).

In this IMO, a total of 22 contestants achieved a perfect score of 42 points.

Problem 1.11 (IMO 39-4, proposed by the United Kingdom).
Determine all pairs (a, b) of positive integers such that $ab^2 + b + 7$ divides $a^2 b + a + b$.

Solution. By the given condition, $ab^2 + b + 7 \mid b(a^2 b + a + b)$. Then $ab^2 + b + 7 \mid a(ab^2 + b + 7) + b^2 - 7a$, and hence $ab^2 + b + 7 \mid b^2 - 7a$.

If $b^2 - 7a > 0$, then $b^2 - 7a \geq ab^2 + b + 7 > b^2$, which is impossible.

If $b^2 - 7a = 0$, then $a = 7k^2$ and $b = 7k$, where k is a positive integer. It is easy to check that every such pair (a, b) satisfies the given condition.

If $b^2 - 7a < 0$, since $ab^2 + b + 7 \leq 7a - b^2 < 7a$, then $b^2 < 7$, so $b = 1$ or 2.

When $b = 1$, we have $a + 8 \mid a^2 + a + 1 = (a + 8)(a - 7) + 57$. Then $a + 8 \mid 57$, from which $a = 11$ or 49.

When $b = 2$, we obtain $4a + 9 \mid 2a^2 + a + 2$. Then

$$4a + 9 \mid 2(2a^2 + a + 2) = a(4a + 9) + 4 - 7a.$$

Hence, $4a + 9 \mid 7a - 4$. However, $7a - 4 < 2(4a + 9)$. Thus, $7a - 4 = 4a + 9$, so $a = \frac{13}{3}$, and this subcase yields no solution.

Summarizing, we see that the solution is

$$(a, b) = (11, 1), (49, 1), (7k^2, 7k), \quad k \in \mathbf{N}^*.$$

Remark. When discussing divisibility problems, two useful methods are factor analysis and scaling. Due to the lack of a good starting point for discussing factors in this problem, and also noting the irregularity of the degrees, we attempted to impose size constraints on a and b, leading to a small number of cases.

【Score Situation】 This particular problem saw the following distribution of scores among contestants: 142 contestants scored 7 points, 23 contestants scored 6 points, 10 contestants scored 5 points, 20 contestants scored 4 points, 38 contestants scored 3 points, 14 contestants scored 2 points, 47 contestants scored 1 point, and 125 contestants scored 0 point. The average score of this problem is 3.463, indicating that it was relatively straightforward.

Among the top five teams in the team scores, the scores of this problem are as follows: The Iran team scored 36 points (with a total team score of 211 points), the Bulgaria team scored 35 points (with a total team score of 195 points), the Hungary team scored 40 points (with a total team score of 186 points), the United States team scored 39 points (with a total team score of 186 points), and the Chinese Taiwan team scored 36 points (with a total team score of 184 points).

The gold medal cutoff for this IMO was set at 31 points (with 37 contestants earning gold medals), the silver medal cutoff was 24 points (with 66 contestants earning silver medals), and the bronze medal cutoff was 14 points (with 102 contestants earning bronze medals).

In this IMO, only one contestant achieved a perfect score of 42 points, namely Omid Amini from Iran.

Problem 1.12 (IMO 43-3, proposed by Romania). Find all pairs of integers $m > 2$ and $n > 2$ such that there are infinitely many positive integers k for which $k^n + k^2 - 1$ divides $k^m + k - 1$.

Solution. First, we prove a lemma.

Lemma. *Suppose $f(x)$ and $g(x)$ are polynomials with rational coefficients and $g(x) \neq 0$. If there exist infinitely many positive integers a such that $\frac{f(a)}{g(a)}$ is an integer, then $f(x)$ is divisible by $g(x)$.*

Proof of the Lemma. Suppose $f(x) = q(x)g(x) + r(x)$, where $q(x)$ and $r(x)$ are both polynomials with rational coefficients and the degree of $r(x)$ is less than the degree of $g(x)$. Let d be the common denominator of the coefficients of $q(x)$. Then $d \cdot q(a)$ is an integer for any integer a.

In this proposition, if $\frac{f(a)}{g(a)} = q(a) + \frac{r(a)}{g(a)}$ is an integer, then $d \cdot \frac{r(a)}{g(a)}$ is an integer. We also observe that

$$\lim_{a \to \infty} d \cdot \frac{r(a)}{g(a)} = 0.$$

Hence there exists a positive integer A such that for $\forall a > A$, both $g(a) \neq 0$ and $\left| d \cdot \frac{r(a)}{g(a)} \right| < 1$ are valid.

Now, if there exist infinitely many positive integers a such that $\frac{f(a)}{g(a)}$ is an integer, then in these integers there exist infinitely many $a > A$ such that $r(a) = 0$. This implies that $r(x) = 0$, which is equivalent to $g \mid f$. The lemma is proved.

Back to the original problem, we observe that

$$\frac{a^5 + a - 1}{a^3 + a^2 - 1} = a^2 - a + 1.$$

Thus, $(m, n) = (5, 3)$ is a solution. We next prove that this is the unique solution.

By the lemma,

$$(x^n + x^2 - 1) \mid (x^m + x - 1).$$

Then $n \leq m$, and

$$(x^n + x^2 - 1) \mid (x + 1)(x^m + x - 1) - (x^n + x^2 - 1).$$

Since

$$(x + 1)(x^m + x - 1) - (x^n + x^2 - 1) = x^n(x^{m-n+1} + x^{m-n} - 1),$$

so

$$(x^n + x^2 - 1) \mid (x^{m-n+1} + x^{m-n} - 1).$$

Then $n \leq m - n + 1$, from which $m - n \geq n - 1 \geq 2$.

Since $0^n + 0^2 - 1 = -1$ and $1^n + 1^2 - 1 = 1$, there exists $\alpha \in (0, 1)$ such that $\alpha^n + \alpha^2 - 1 = 0$. Furthermore, $\alpha^{m-n+1} + \alpha^{m-n} - 1 = 0$. However, from $n \leq m - n + 1$ and $m - n \geq 2$ we have $\alpha^n \geq \alpha^{m-n+1}$ and $\alpha^2 \geq \alpha^{m-n}$. Thus, $\alpha^{m-n+1} + \alpha^{m-n} - 1 \leq \alpha^n + \alpha^2 - 1 = 0$. Hence, the above inequalities should actually be equalities, which means $n = m - n + 1$ and $m - n = 2$. The solution is $(m, n) = (5, 3)$.

Remark. The lemma is relatively common and not difficult to prove. Compared to the lemma, the rest part of the solution involves an estimation of degrees and discussion on zeros of polynomials, which may actually be more difficult. This solution implies that number theory problems may not necessarily be limited to the range of integers. When it comes to polynomials, examining zeros in the range of real numbers can be helpful.

【Score Situation】This particular problem saw the following distribution of scores among contestants: 14 contestants scored 7 points, 2 contestants scored 6 points, 4 contestants scored 5 points, no contestant scored 4 points, 2 contestants scored 3 points, 1 contestant scored 2 points, 145 contestants scored 1 point, and 311 contestants scored 0 point. The average score of this problem is 0.591, indicating that it was extremely difficult.

Among the top five teams in the team scores, the scores of this problem are as follows: The China team scored 24 points (with a total team score of 212 points), the Russia team scored 16 points (with a total team score of 204 points), the United States team scored 10 points (with a total team score of 171 points), the Bulgaria team scored 11 points (with a total team score of 167 points), and the Vietnam team scored 18 points (with a total team score of 166 points).

The gold medal cutoff for this IMO was set at 29 points (with 39 contestants earning gold medals), the silver medal cutoff was 23 points (with 73 contestants earning silver medals), and the bronze medal cutoff was 14 points (with 120 contestants earning bronze medals).

In this IMO, only three contestants achieved a perfect score of 42 points, namely Yunhao Fu and Botong Wang from China, and Andrei Khaliavine from Russia.

Problem 1.13 (IMO 43-4, proposed by Romania). The positive divisors of an integer $n > 1$ are $d_1 < d_2 < \cdots < d_k$, so that $d_1 = 1$ and $d_k = n$. Let $D = d_1 d_2 + d_2 d_3 + \cdots + d_{k-1} d_k$. Show that $D < n^2$ and find all n for which D divides n^2.

Proof. We observe that if d is a divisor of n, then $\frac{n}{d}$ is also a divisor of n. Thus for all $1 \leq i \leq k$, we have $d_i d_{k+1-i} = n$. Therefore,

$$D = d_1 d_2 + d_2 d_3 + \cdots + d_{k-1} d_k$$

$$= \frac{n^2}{d_k d_{k-1}} + \frac{n^2}{d_{k-1} d_{k-2}} + \cdots + \frac{n^2}{d_2 d_1}$$

$$= n^2 \left(\frac{1}{d_1 d_2} + \frac{1}{d_2 d_3} + \cdots + \frac{1}{d_{k-1} d_k} \right)$$

$$\leq n^2 \left[\left(\frac{1}{d_1} - \frac{1}{d_2} \right) + \left(\frac{1}{d_2} - \frac{1}{d_3} \right) + \cdots + \left(\frac{1}{d_{k-1}} - \frac{1}{d_k} \right) \right]$$

$$= n^2 \left(\frac{1}{d_1} - \frac{1}{d_k} \right)$$

$$< \frac{n^2}{d_1} = n^2.$$

To find all n for which D divides n^2, let p be the smallest prime factor of n. Then $d_2 = p$ and $d_{k-1} = \frac{n}{p}$. If $n = p$, then $k = 2$, $D = p = n$, and $D \,|\, n^2$. If $n \neq p$, then n is a composite number with $k > 2$. We have $D > d_{k-1} d_k = \frac{n}{p} \cdot n = \frac{n^2}{p}$. If $D \,|\, n^2$, then $\frac{n^2}{D}$ is also a divisor of n^2. However, $1 < \frac{n^2}{D} < p$, which is a contradiction to the fact that p is the smallest prime factor of n and n^2.

Consequently, D divides n^2 if and only if n is a prime.

【Score Situation】 This particular problem saw the following distribution of scores among contestants: 176 contestants scored 7 points, 20 contestants scored 6 points, 9 contestants scored 5 points, 70 contestants scored 4 points, 38 contestants scored 3 points, 15 contestants scored 2 points, 45 contestants scored 1 point, and 106 contestants scored 0 point. The average score of this problem is 3.896, indicating that it was relatively straightforward.

Among the top five teams in the team scores, the scores of this problem are as follows: The China team scored 42 points (with a total team score of 212 points), the Russia team scored 42 points (with a total team score of 204 points), the United States team scored 24 points (with a total team score of 171 points), the Bulgaria team scored 34 points (with a total team score of 167 points), and the Vietnam team scored 35 points (with a total team score of 166 points).

The gold medal cutoff for this IMO was set at 29 points (with 39 contestants earning gold medals), the silver medal cutoff was 23 points (with 73 contestants earning silver medals), and the bronze medal cutoff was 14 points (with 120 contestants earning bronze medals).

In this IMO, only three contestants achieved a perfect score of 42 points, namely Yunhao Fu and Botong Wang from China, and Andrei Khaliavine from Russia.

Problem 1.14 (IMO 50-1, proposed by Australia). Let n be a positive integer and let $a_1, a_2, \ldots, a_k (k \geq 2)$ be distinct integers in the set

$\{1, \ldots, n\}$ such that n divides $a_i(a_{i+1} - 1)$ for $i = 1, \ldots, k - 1$. Prove that n does not divide $a_k(a_1 - 1)$.

Proof. We first prove by induction that $n \mid a_1(a_m - 1)$ for any integer $2 \leq m \leq k$.

If $m = 2$, we have $n \mid a_1(a_2 - 1)$ by assumption. Suppose that $n \mid a_1(a_m - 1)$ for some $2 \leq m \leq k - 1$; then

$$n \mid a_1(a_m - 1)(a_{m+1} - 1).$$

By assumption, $n \mid a_m (a_{m+1} - 1)$, so $n \mid a_1 a_m (a_{m+1} - 1)$, and therefore

$$n \mid (a_1 a_m(a_{m+1} - 1) - a_1(a_m - 1)(a_{m+1} - 1)),$$

i.e., $n \mid a_1(a_{m+1} - 1)$. By the method of induction, $n \mid a_1(a_m - 1)$ holds for every integer $2 \leq m \leq k$. In particular, $n \mid a_1(a_k - 1)$.

Since $a_1(a_k - 1) - a_k(a_1 - 1) = a_k - a_1$ is not divisible by n (because a_1 and a_k are distinct elements in $\{1, \ldots, n\}$), we conclude that $a_k(a_1 - 1)$ is not divisible by n.

【Score Situation】 This particular problem saw the following distribution of scores among contestants: 324 contestants scored 7 points, 21 contestants scored 6 points, 16 contestants scored 5 points, 17 contestants scored 4 points, 20 contestants scored 3 points, 28 contestants scored 2 points, 56 contestants scored 1 point, and 83 contestants scored 0 point. The average score of this problem is 4.804, indicating that it was simple.

Among the top five teams in the team scores, the scores of this problem are as follows: The China team scored 42 points (with a total team score of 221 points), the Japan team scored 42 points (with a total team score of 212 points), the Russia team scored 42 points (with a total team score of 203 points), the South Korea team scored 42 points (with a total team score of 188 points), and the North Korea team scored 42 points (with a total team score of 183 points).

The gold medal cutoff for this IMO was set at 32 points (with 49 contestants earning gold medals), the silver medal cutoff was 24 points (with 98 contestants earning silver medals), and the bronze medal cutoff was 14 points (with 135 contestants earning bronze medals).

In this IMO, only two contestants achieved a perfect score of 42 points, namely Dongyi Wei from China and Makoto Soejima from Japan.

Problem 1.15 (IMO 64-1, proposed by Colombia). Determine all composite integers $n > 1$ that satisfy the following property: if $d_1, d_2, \ldots, d_k$ are all the positive divisors of n with $l = d_1 < d_2 < \cdots < d_k = n$, then d_i divides $d_{i+1} + d_{i+2}$ for every $1 \leq i \leq k - 2$.

Solution. All composite integers n satisfying the given property are those $n = p^\alpha$, where p is a prime number and integer $\alpha \geq 2$.

First, for such n, we have $d_i = p^{i-1}$ $(i = 1, 2, \ldots, \alpha+1)$. Then evidently $d_i \mid d_{i+1} + d_{i+2}$. Otherwise we assume that $p < q$ are the smallest two prime factors of n. Suppose that for some positive integer t, there is $p^t < q < p^{t+1}$. Then $d_j = p^{j-1}$ for $j = 1, 2, \ldots, t+1$ and $d_{t+2} = q$.

If $t \geq 2$, then $d_t = p^{t-1}$, $d_{t+1} = p^t$, and $d_{t+2} = q$, so $d_{t+1} + d_{t+2}$ is not divisible by d_t, a contradiction. Therefore $t = 1$. Now, $d_2 = p$ and $d_3 = q$, so $d_{k-2} = \frac{n}{q}$ and $d_{k-1} = \frac{n}{p}$. Since $d_{k-2} \mid d_{k-1} + d_k$, we have $\frac{n}{q} \mid \frac{n}{p} + n$. Hence, $p \mid q + pq$, also a contradiction. Summarizing, n does not have two distinct prime factors and must be p^α, where p is a prime number and integer $\alpha \geq 2$.

Remark. Similar to Problem 1.13, we considered those pairs of factors $\left(p, \frac{n}{p}\right)$ and $\left(q, \frac{n}{q}\right)$ to obtain the contradiction. In fact the key point is to find the form of n satisfying the given condition.

【Score Situation】 This particular problem saw the following distribution of scores among contestants: 474 contestants scored 7 points, 8 contestants scored 6 points, 6 contestants scored 5 points, 9 contestants scored 4 points, 9 contestants scored 3 points, 67 contestants scored 2 points, 19 contestants scored 1 point, and 26 contestants scored 0 point. The average score of this problem is 5.845, indicating that it was simple.

Among the top five teams in the team scores, the scores of this problem are as follows: The China team scored 42 points (with a total team score of 240 points), the United States team scored 42 points (with a total team score of 222 points), the South Korea team scored 42 points (with a total team score of 215 points), the Romania team scored 42 points (with a total team score of 208 points), and the Canada team scored 42 points (with a total team score of 183 points).

The gold medal cutoff for this IMO was set at 32 points (with 54 contestants earning gold medals), the silver medal cutoff was 25 points (with 90 contestants earning silver medals), and the bronze medal cutoff was 18 points (with 170 contestants earning bronze medals).

In this IMO, a total of five contestants achieved a perfect score of 42 points.

1.2.2 *Prime numbers, prime factors, and coprime numbers*

Problem 1.16 (IMO 11-1, proposed by the German Democratic Republic). Prove that there are infinitely many natural numbers a with the following property: the number $z = n^4 + a$ is not prime for any natural number n.

Proof. Let $a = 4m^4$, where m is a positive integer greater than 1. Then

$$n^4 + a = n^4 + 4m^4$$

$$= \left(n^2 + 2m^2\right)^2 - 4m^2n^2$$

$$= (n^2 + 2m^2 + 2mn)(n^2 + 2m^2 - 2mn).$$

Note that

$$n^2 + 2m^2 + 2mn > n^2 + 2m^2 - 2mn$$

$$= (n-m)^2 + m^2 > 1,$$

and hence $z = n^4 + a$ can be written as the product of two positive integers greater than 1. So z is not prime, and there are infinitely many a with the required property.

Remark. The key to this problem is the factorization of $x^4 + 4y^4$. For those who are not familiar with this factorization, the form of a can be decided by the method of undetermined coefficients. Specifically, if we want that $x^4 + a = (x^2 + sx + t)(x^2 + s'x + t')$, then it should be $s + s' = 0$, $t = t'$, $s^2 = 2t$, and $t^2 = a$. Now we let $s = -s' = 2m$ and $t = t' = 2m^2$. Then $a = 4m^4$.

【Score Situation】 This particular problem saw the following distribution of scores among contestants: 45 contestants scored 5 points, 3 contestants scored 4 points, 6 contestants scored 3 points, 7 contestants scored 2 points, 17 contestants scored 1 point, and 34 contestants scored 0 point. The average score of this problem is 2.544, indicating that it had a certain level of difficulty.

Among the top five teams in the team scores, the scores of this problem are as follows: The Hungary team scored 28 points (with a total team score of 247 points), the German Democratic Republic team scored 38 points (with a total team score of 240 points), the Soviet Union team scored 30 points (with a total team score of 231 points), the Romania team scored 33 points (with a total team score of 219 points), and the United Kingdom team scored 20 points (with a total team score of 193 points).

The gold medal cutoff for this IMO was set at 40 points (with 3 contestants earning gold medals), the silver medal cutoff was 30 points (with 20 contestants earning silver medals), and the bronze medal cutoff was 24 points (with 21 contestants earning bronze medals).

In this IMO, only three contestants achieved a perfect score of 40 points, namely Tibor Fiala from Hungary, Vladimir Drinfeld from the Soviet Union and Simon Phillips Norton from the United Kingdom.

Problem 1.17 (IMO 24-3, proposed by Germany). Let a, b, and c be positive integers, no two of which have a common divisor greater than 1. Show that $2abc - ab - bc - ca$ is the largest integer which cannot be expressed in the form $xbc + yca + zab$, where x, y, and z are non-negative integers.

Proof 1. First, we prove that $2abc - ab - bc - ca$ cannot be expressed in the form $xbc + yca + zab$. Assume that non-negative integers $x, y, z \geq 0$ satisfy the expression $2abc - ab - bc - ca = xbc + yca + zab$. Then

$$2abc = bc\,(x + 1) + ca\,(y + 1) + ab(z + 1),$$

from which $a \mid bc\,(x + 1)$. Since $(a, bc) = 1$, then $a \mid x+1$, and since $x+1 > 0$, we have $x + 1 \geq a$. Similarly, $y + 1 \geq b$ and $z + 1 \geq c$. This leads to the contradiction that

$$2abc = bc\,(x + 1) + ca\,(y + 1) + ab\,(z + 1) \geq 3abc.$$

Now we show that for every positive integer $n > 2abc - ab - bc - ca$, it is expressible in the form $xbc + yca + zab$, where x, y, z are non-negative integers.

As $(ab, c) = 1$, by Bézout's theorem, the equation $abz + cw = n$ has a solution (z, w), where z and w are integers. Then the pairs of integers in the form $(z + c, w - ab)$ and $(z - c, w + ab)$ are also solutions. So we can choose an integer z such that $0 \leq z < c$.

Now,

$$cw = n - abz$$

$$> 2abc - ab - bc - ca - abz$$

$$\geq 2abc - ab - bc - ca - ab\,(c - 1)$$

$$= abc - bc - ca,$$

and thus $w > ab - a - b$.

As $(a, b) = 1$, by Bézout's theorem, the equation $ay + bx = w$ has a solution (x, y), where x and y are integers. Similarly, we can choose y in the form $(x + a, y - b)$ and $(x - a, y + b)$ such that $0 \leq y < b$.

Now,

$$bx = w - ay$$

$$> ab - a - b - ay$$

$$\geq ab - a - b - a\,(b - 1)$$

$$= -b.$$

Thus, $bx > -b$, and we know that $x > -1$ is an non-negative integer. Then x, y, z are non-negative integers satisfying

$$xbc + yca + zab = (xb + ya)\,c + zab = abz + cw = n.$$

Finally, we conclude that $2abc - ab - bc - ca$ is the largest integer which cannot be expressed in the form $xbc + yca + zab$, where x, y, and z are non-negative integers.

Proof 2. We use the same method to prove that $2abc - ab - bc - ca$ cannot be expressed in the form $xbc + yca + zab$. However we will prove the other half of the problem by a new method.

First,

$$2abc - ab - bc - ca = abc\left(2 - \frac{1}{a} - \frac{1}{b} - \frac{1}{c}\right) \geq abc\left(2 - \frac{1}{1} - \frac{1}{2} - \frac{1}{3}\right) > 0.$$

If $n \equiv 0 \,(\mathrm{mod}\ abc)$, then we can suppose $n = tabc$, where t is a non-negative integer. Let $x = at$ and $y = z = 0$.

If $n \not\equiv 0 \,(\mathrm{mod}\ abc)$, since $(bc, a) = 1$, $(ca, b) = 1$, and $(ab, c) = 1$, then there exsit $0 < x_0 < a$, $0 < y_0 < b$, and $0 < z_0 < c$, such that

$$bcx_0 \equiv n \,(\mathrm{mod}\ a),$$

$$cay_0 \equiv n \,(\mathrm{mod}\ b),$$

$$abz_0 \equiv n \,(\mathrm{mod}\ c).$$

Now, let $m = x_0bc + y_0ca + z_0ab$. Since no two of a, b, c have a common divisor greater than 1, so $m \equiv 0 \,(\mathrm{mod}\ abc)$, and thus we can suppose $n = m + kabc$, where k is an integer. Observe that

$$n > 2abc - ab - bc - ca,$$

$$m \leq bc(a-1) + ca(b-1) + ab(c-1) = 3abc - ab - bc - ca.$$

Hence, $k \geq 0$. Then let $x = x_0 + ka$, $y = y_0$, and $z = z_0$. We have $n = xbc + yca + zab$, with non-negative integers x, y, and z.

Remark. These two proofs are from two natural ideas. In the first proof, we started from a simpler case of two numbers, then we used the same approach to complete the proof for three numbers in the original problem. If we want, this conclusion can be extended to the case of n numbers. The second proof uses congruence methods, which will be the main focus of the next chapter. In this proof, we first found the condition x, y, z

must satisfy, modulo a, b, c. It is natural for us to select the smallest non-negative integers satisfying this congruence condition, then we make some adjustments to obtain the desired n.

【Score Situation】 This particular problem saw the following distribution of scores among contestants: 12 contestants scored 7 points, 2 contestants scored 6 points, 2 contestants scored 5 points, 8 contestants scored 4 points, no contestant scored 3 points, 2 contestants scored 2 points, 1 contestant scored 1 point, and 13 contestants scored 0 point. The average score of this problem is 3.575, indicating that it was relatively straightforward.

Among the top five teams in the team scores, the Germany team achieved a total score of 212 points, the United States team achieved a total score of 171 points, the Hungary team achieved a total score of 170 points, the Soviet Union team achieved a total score of 169 points, and the Romania team achieved a total score of 161 points.

The gold medal cutoff for this IMO was set at 38 points (with 9 contestants earning gold medals), the silver medal cutoff was 26 points (with 27 contestants earning silver medals), and the bronze medal cutoff was 15 points (with 57 contestants earning bronze medals).

In this IMO, a total of four contestants achieved a perfect score of 42 points.

Problem 1.18 (IMO 32-2, proposed by Romania). Let $n > 6$ be an integer and $a_1, a_2, \ldots, a_k$ be all the natural numbers less than n and relatively prime to n. If $a_2 - a_1 = a_3 - a_2 = \cdots = a_k - a_{k-1} > 0$, prove that n must be either a prime number or a power of 2.

Proof 1. Evidently, $a_1 = 1$, $a_k = n - 1$. Let $a_2 - a_1 = a_3 - a_2 = \cdots = a_k - a_{k-1} = d$.

If $a_2 = 2$, then $d = 1$. This means that n is relatively prime to all the positive integers less than n, and hence n is a prime number.

If $a_2 = 3$, then $d = 2$. This means that n is relatively prime to all the positive odd numbers less than n, and hence n is a power of 2.

If $a_2 > 3$, then $a_2 = 1 + d$ is the least prime number that cannot divide n. We have $2 \mid n$ and $3 \mid n$. Since $n - 1 = a_k = 1 + (k - 1)d$, so

$$(k - 1)d = n - 2 \equiv 1 \pmod{3},$$

and hence $(3, d) = 1$. On the other hand, $(3, 1 + d) = (3, a_2) = 1$, so $d \equiv 1 \pmod 3$ and $3 \mid 1 + 2d$.

If $1 + 2d < n$, then $(a_3, n) = (1 + 2d, n) \geq 3$, and we have a contradiction.

If $1 + 2d \geq n$, then in all the positive integers less than n, there are only two numbers, 1 and $n - 1$, relatively prime to n. Now suppose $n = 2m$ ($m > 3$). If m is even, then $(m + 1, n) = 1$; if m is odd, then $(m + 2, n) = 1$, a contradiction.

As a conclusion, a suitable n must be either a prime number or a power of 2.

Proof 2. Again let $a_2 - a_1 = a_3 - a_2 = \cdots = a_k - a_{k-1} = d$. We discuss in two conditions.

If n has a factor which is a perfect square, say $n = a^2 \cdot b$ $(a > 1)$, then $(ab - 1, n) = 1$ and $(ab + 1, n) = 1$. Since $1 \le ab - 1 < ab + 1 < n$ and $(ab, n) = ab > 1$, so $ab - 1$ and $ab + 1$ are two adjacent terms which are less than n and relatively prime to n. Thus, $d = (ab + 1) - (ab - 1) = 2$. Now that $a_1 = 1$, and then n is relatively prime to all the positive odd numbers less than n. Hence n is a power of 2.

Now we discuss the case if n does not have such a factor which is a perfect square.

If n is odd, then $a_1 = 1$ and $a_2 = 2$, so $d = 1$. And n is relatively prime to all the positive integers less than n, from which n is a prime number.

If n is even, then suppose $n = 2m(m > 3)$. Since n does not have such a factor which is a perfect square, m is odd. Then $(m - 2, 2m) = 1$, $(m + 2, 2m) = 1$, $(m - 1, 2m) = 2$, $(m + 1, 2m) = 2$, and $(m, 2m) = m$. Since $1 < m - 2 < m + 2 < 2m = n$, so $m - 2$ and $m + 2$ are two adjacent terms which are less than n and relatively prime to n. Thus $d = (m + 2) - (m - 2) = 4$. Now that $a_1 = 1$, $a_2 = 5$, and $a_k = n - 1 > 5$. Then there exists a_3 and $a_3 = 9$. However, for $a_2 = 5$ we know n is not divisible by 3, which is a contradiction to $(a_3, n) = (9, n) \ge 3$.

As a conclusion, a suitable n must be either a prime number or a power of 2.

Remark. We proved the problem by two methods from different perspectives a and n, respectively. The analysis of the common difference in the first proof is relatively straightforward while the discussion in the second proof on whether n has a perfect square factor seems a little special. However, both proofs originate from a common problem-solving method, which is to find clues from simple cases.

【Score Situation】 This particular problem saw the following distribution of scores among contestants: 130 contestants scored 7 points, 16 contestants scored 6 points, 6 contestants scored 5 points, 31 contestants scored 4 points, 27 contestants scored 3 points, 21 contestants scored 2 points, 34 contestants scored 1 point, and 47 contestants scored 0 point. The average score of this problem is 4.221, indicating that it was simple.

Among the top five teams in the team scores, the scores of this problem are as follows: The Soviet Union team scored 42 points (with a total team score of 241 points), the China team scored 42 points (with a total team score of 231 points), the Romania team scored 42 points (with a total team score of 225 points), the Germany team scored 41 points (with a total team score of 222 points), and the United States team scored 42 points (with a total team score of 212 points).

The gold medal cutoff for this IMO was set at 39 points (with 20 contestants earning gold medals), the silver medal cutoff was 31 points (with 51 contestants earning silver medals), and the bronze medal cutoff was 19 points (with 84 contestants earning bronze medals).

In this IMO, a total of nine contestants achieved a perfect score of 42 points.

Problem 1.19 (IMO 35-6, proposed by Finland). Show that there exists a set A of positive integers with the following property: For any infinite set S of primes there exist two positive integers $m \in A$ and $n \notin A$ each of which is a product of k distinct elements of S for some $k \geq 2$.

Proof. Let all prime numbers be $p_1 = 2, p_2 = 3, p_3 = 5, \ldots$, successively. For $i = 1, 2, \ldots$, let

$$A_i = \{p_i p_{j_1} p_{j_2} \cdots p_{j_i} \mid i < j_1 < j_2 < \cdots < j_i\},$$

$$A = \bigcup_{i=1}^{\infty} A_i.$$

For any infinite set S of primes, suppose the prime numbers in S are $p_{s_1} < p_{s_2} < \cdots$, successively. Then letting $k = s_1 + 1$, we have

$$m = p_{s_1} p_{s_2} \cdots p_{s_k} \in A_{s_1} \subset A,$$

$$n = p_{s_2} p_{s_3} \cdots p_{s_{k+1}} \notin A.$$

Hence, the set A satisfies the given condition.

Remark. This proof seems concise, however the construction itself is not easy to find because S is highly uncertain. We note that no matter S contains what primes, there always exists a unique least prime number. Based on this important feature, S could be classified, then we reached this proof.

【Score Situation】This particular problem saw the following distribution of scores among contestants: 88 contestants scored 7 points, 3 contestants scored 6 points, 8 contestants scored 5 points, 4 contestants scored 4 points, 7 contestants scored 3 points, 30 contestants scored 2 points, 34 contestants scored 1 point, and 211 contestants scored 0 point. The average score of this problem is 2.091, indicating that it had a certain level of difficulty.

Among the top five teams in the team scores, the scores of this problem are as follows: The United States team scored 42 points (with a total team score of 252 points), the China team scored 21 points (with a total team score of 229 points), the Russia team scored 31 points (with a total team score of 224 points), the Bulgaria team scored 28 points (with a total team score of 223 points), and the Hungary team scored 34 points (with a total team score of 221 points).

The gold medal cutoff for this IMO was set at 40 points (with 30 contestants earning gold medals), the silver medal cutoff was 30 points (with 64 contestants earning silver medals), and the bronze medal cutoff was 19 points (with 98 contestants earning bronze medals).

In this IMO, a total of 22 contestants achieved a perfect score of 42 points.

Problem 1.20 (IMO 39-3, proposed by Byelorussia). For any positive integer n, let $d(n)$ denote the number of positive divisors of n (including 1 and n itself). Determine all positive integers k such that $\frac{d(n^2)}{d(n)} = k$ for some n.

Solution. When $n = 1$, we have $\frac{d(n^2)}{d(n)} = 1$, so $k = 1$ is one of the solutions.

When $n > 1$, we factor n into the product of prime numbers: $n = p_1^{\alpha_1} p_2^{\alpha_2} \cdots p_m^{\alpha_m}$. Then

$$\frac{d(n^2)}{d(n)} = \frac{(2\alpha_1 + 1)(2\alpha_2 + 1)\cdots(2\alpha_m + 1)}{(\alpha_1 + 1)(\alpha_2 + 1)\cdots(\alpha_m + 1)}$$

$$= \frac{(2\beta_1 - 1)(2\beta_2 - 1)\cdots(2\beta_m - 1)}{\beta_1 \beta_2 \cdots \beta_m},$$

where $\beta_j = \alpha_j + 1$, $j = 1, 2, \ldots, m$.

Therefore, the problem is equivalent to determining all positive integers k that can be expressed in the form:

$$k = \frac{(2\beta_1 - 1)(2\beta_2 - 1)\cdots(2\beta_m - 1)}{\beta_1 \beta_2 \cdots \beta_m},$$

where m is a positive integer, β_j $(j = 1, 2, \ldots, m)$ are positive integers greater than 1. We define these integers k as "expressible." Evidently, the "expressible" integers must be odd. Now by mathematical induction we will prove that all the odd numbers greater than 1 are "expressible".

First, $3 = \frac{9}{5} \times \frac{5}{3} = \frac{(2 \times 5 - 1)(2 \times 3 - 1)}{5 \times 3}$, so 3 is "expressible".

For every positive odd number, there exists one and only one way to express the number as $2^{t+1}s + 2^t - 1$, where s is a non-negative integer and t is a positive integer (we can write the number in binary. Then from right to left, all the consecutive 1s together form $2^t - 1$, and the rest is $2^{t+1}s$). Now assume all the positive odd numbers less than $2^{t+1}s + 2^t - 1$ are "expressible," and we will prove that $2^{t+1}s + 2^t - 1$ is also "expressible"

Let $\beta_1 = (2^t - 1)(2s + 1)$ and $\beta_{j+1} = 2\beta_j - 1$, $j = 1, 2, \ldots, t - 1$. Then

$$\beta_t - 1 = 2^{t-1}(\beta_1 - 1),$$

$$2\beta_t - 1 = 2^t (\beta_1 - 1) + 1$$

$$= 2^t (2^t - 1) (2s + 1) - 2^t + 1$$

$$= (2^t - 1) (2^{t+1}s + 2^t - 1).$$

Hence,

$$A = \frac{(2\beta_1 - 1)(2\beta_2 - 1) \cdots (2\beta_t - 1)}{\beta_1 \beta_2 \cdots \beta_t}$$

$$= \frac{2\beta_t - 1}{\beta_1}$$

$$= \frac{2^{t+1}s + 2^t - 1}{2s + 1}.$$

If $s = 0$, then $A = 2^{t+1}s + 2^t - 1$ is "expressible." Otherwise, $2s + 1 < 2^{t+1}s + 2^t - 1$, and with the assumption of the induction hypothesis we know $2s+1$ is "expressible." Suppose it can be expressed as B. Then $2^{t+1}s + 2^t - 1$ can be expressed as AB. Hence by mathematical induction we know that all the odd numbers greater than 1 are "expressible." The solutions are all the odd numbers.

Remark. The formula for the number (or sum) of positive factors from the prime factorization is useful. In this problem, the form $2^{t+1}s + 2^t - 1$ seems tricky. Actually, it is derived from the desire to reduce the fraction, i.e., to make $2\beta_j - 1$ equal to β_{j+1}. This leads to $2\beta_t - 1 = 2^t(\beta_1 - 1) + 1$, and together with the assumption of the induction hypothesis we wanted to find a relatively small positive integer r such that $r \cdot (2^t(\beta_1 - 1) + 1) = n \cdot \beta_1$. Then we got the form $2^{t+1}s + 2^t - 1$.

【Score Situation】 This particular problem saw the following distribution of scores among contestants: 30 contestants scored 7 points, 5 contestants scored 6 points, 1 contestant scored 5 points, 3 contestants scored 4 points, 20 contestants scored 3 points, 152 contestants scored 2 points, 117 contestants scored 1 point, and 91 contestants scored 0 point. The average score of this problem is 1.761, indicating that it was relatively challenging.

Among the top five teams in the team scores, the scores of this problem are as follows: The Iran team scored 18 points (with a total team score of 211 points), the Bulgaria team scored 15 points (with a total team score of 195 points), the Hungary team scored 25 points (with a total team score of 186 points), the United States team scored 34 points (with a total team score of 186 points), and the Chinese Taiwan team scored 21 points (with a total team score of 184 points).

The gold medal cutoff for this IMO was set at 31 points (with 37 contestants earning gold medals), the silver medal cutoff was 24 points (with 66 contestants earning silver medals), and the bronze medal cutoff was 14 points (with 102 contestants earning bronze medals).

In this IMO, only one contestant achieved a perfect score of 42 points, namely Omid Amini from Iran.

Problem 1.21 (IMO 42-6, proposed by Bulgaria). Let a, b, c, d be integers with $a > b > c > d > 0$. Suppose that

$$ac + bd = (b + d + a - c)(b + d - a + c).$$

Prove that $ab + cd$ is not prime.

Proof 1. By the given condition,

$$ac + bd = (b + d)^2 - (a - c)^2 = b^2 + d^2 - (a^2 + c^2) + 2(ac + bd),$$

so $a^2 - ac + c^2 = b^2 + bd + d^2$, and then

$$(ac + bd)(b^2 + bd + d^2) = ac(b^2 + bd + d^2) + bd(a^2 - ac + c^2)$$
$$= (ab + cd)(ad + bc).$$

Assume $ab + cd$ is prime. Since

$$ab + cd - (ac + bd) = (a - d)(b - c) > 0,$$

we know $ac + bd < ab + cd$. Thus $ac + bd$ is relatively prime to $ab + cd$, thus we must have $ac + bd \mid ad + bc$. However,

$$ac + bd - (ad + bc) = (a - b)(c - d) > 0,$$

and so $ac + bd > ad + bc$, which is a contradiction. As a result, $ab + cd$ is not prime.

Proof 2. Assume $ab + cd$ is prime. Since

$$ab + cd = (a + d)c + (b - c)a = m(a + d, b - c)$$

for some $m \in \mathbf{N}^*$, we see that $m = 1$ or $(a + d, b - c) = 1$.

If $m = 1$, then

$$(a + d, b - c) = ab + cd$$

$$> ab + cd - (a - b + c + d)$$

$$= (a + d)(c - 1) + (b - c)(a + 1)$$

$$\geq (a + d, b - c),$$

which is a contradiction.

If $(a + d, b - c) = 1$, then since

$$(a + d)b - (b - c)a = ac + bd = (b + d + a - c)(b + d - a + c),$$

we know

$$(a + d)(a - c - d) = (b - c)(b + d + c).$$

Therefore, $a - c - d > 0$, and there exists $k \in \mathbf{N}^*$ such that

$$a - c - d = k(b - c),$$

$$b + d + c = k(a + d).$$

By adding these two equalities together, we have

$$a + b = k(a + b - c + d),$$

i.e., $\qquad k(c - d) = (k - 1)(a + b).$

If $k = 1$, then $c = d$, a contradiction. If $k > 1$, then

$$2 \geq \frac{k}{k - 1} = \frac{a + b}{c - d} > \frac{a + b}{c} > 2,$$

also a contradiction. Summarizing, we see that $ab + cd$ is not prime.

Proof 3. By $ac + bd = (b + d + a - c)(b + d - a + c) = (b + d)^2 - (a - c)^2$, we know

$$a^2 - ac + c^2 = b^2 + bd + d^2.$$

So, we can construct a convex quadrilateral $ABCD$, such that $AB = a$, $AD = c$, $CB = d$, $CD = b$, and $\angle BAD = 60°$ and $\angle BCD = 120°$. Then

$ABCD$ is a cyclic quadrilateral, and

$$BD^2 = a^2 - ac + c^2. \tag{1}$$

Let $\angle ABC = \alpha$. Then $\angle ADC = 180° - \alpha$. For $\triangle ABC$ and $\triangle ADC$, by the law of cosines,

$$AC^2 = a^2 + d^2 - 2ad \cos \alpha = b^2 + c^2 - 2bc \cos(180° - \alpha)$$

$$= b^2 + c^2 + 2bc \cos \alpha,$$

which leads to $2 \cos \alpha = \frac{a^2 + d^2 - b^2 - c^2}{ad + bc}$. Therefore,

$$AC^2 = a^2 + d^2 - ad \cdot \frac{a^2 + d^2 - b^2 - c^2}{ad + bc}$$

$$= \frac{(a^2 + d^2)(ad + bc) - ad(a^2 + d^2 - b^2 - c^2)}{ad + bc}$$

$$= \frac{(a^2 + d^2) bc + ad(b^2 + c^2)}{ad + bc}$$

$$= \frac{(ab + cd)(ac + bd)}{ad + bc}. \tag{2}$$

In cyclic quadrilateral $ABCD$, by Ptolemy's theorem, $AC \cdot BD = a \cdot b + c \cdot d$, i.e.,

$$AC^2 \cdot BD^2 = (ab + cd)^2. \tag{3}$$

Substituting equality (1) and equality (2) into equality (3), we reach

$$(a^2 - ac + c^2)(ac + bd) = (ab + cd)(ad + bc).$$

Hence,

$$ac + bd \mid (ab + cd)(ad + bc).$$

The last part of the proof is the same as in the first proof.

Remark. We proved the conclusion by contradiction in three ways. It is worth mentioning that two solutions (a, b, c, d) satisfying the given condition are $(21, 18, 14, 1)$ and $(65, 50, 34, 11)$.

Moreover, in the third proof, we constructed a cyclic quadrilateral $ABCD$ to help us. We can find the following conclusions: if positive integers

a, b, c, d satisfy $a^2 - ac + c^2 = b^2 + bd + d^2 = M$, then

$$\frac{(ab + cd)(ad + bc)}{ac + bd} = M.$$

Also we can observe that if $ac + bd \neq 0$, then

$$\frac{(ab + cd)(ad + bc)}{ac + bd} = \frac{(a^2 + c^2)bd + (b^2 + d^2)ac}{ac + bd}$$

$$= \frac{(a^2 - ac + c^2)bd + (b^2 + bd + d^2)ac}{ac + bd}$$

$$= \frac{Mbd + Mac}{ac + bd} = M.$$

【Score Situation】 This particular problem saw the following distribution of scores among contestants: 27 contestants scored 7 points, 3 contestants scored 6 points, 9 contestants scored 5 points, 10 contestants scored 4 points, 8 contestants scored 3 points, 16 contestants scored 2 points, 20 contestants scored 1 point, and 380 contestants scored 0 point. The average score of this problem is 0.778, indicating that it was extremely difficult.

Among the top five teams in the team scores, the scores of this problem are as follows: The China team scored 36 points (with a total team score of 225 points), the Russia team scored 21 points (with a total team score of 196 points), the United States team scored 21 points (with a total team score of 196 points), the Bulgaria team scored 36 points (with a total team score of 185 points), the South Korea team scored 11 points (with a total team score of 185 points).

The gold medal cutoff for this IMO was set at 30 points (with 39 contestants earning gold medals), the silver medal cutoff was 20 points (with 81 contestants earning silver medals), and the bronze medal cutoff was 11 points (with 122 contestants earning bronze medals).

In this IMO, a total of four contestants achieved a perfect score of 42 points.

Problem 1.22 (IMO 59-5, proposed by Mongolia). Let $a_1, a_2, \ldots$ be an infinite sequence of positive integers. Suppose that there is an integer $N > 1$ such that, for each $n \geq N$, the number

$$\frac{a_1}{a_2} + \frac{a_2}{a_3} + \cdots + \frac{a_{n-1}}{a_n} + \frac{a_n}{a_1}$$

is an integer. Prove that there is a positive integer M such that $a_m = a_{m+1}$ for all $m \geq M$.

Proof. By the given condition, for integer $n \geq N$,

$$\left(\frac{a_1}{a_2} + \frac{a_2}{a_3} + \cdots + \frac{a_n}{a_{n+1}} + \frac{a_{n+1}}{a_1} \right) - \left(\frac{a_1}{a_2} + \frac{a_2}{a_3} + \cdots + \frac{a_{n-1}}{a_n} + \frac{a_n}{a_1} \right)$$

$$= \frac{a_n}{a_{n+1}} + \frac{a_{n+1}}{a_1} - \frac{a_n}{a_1}$$

is an integer. Then $\frac{a_1 a_n}{a_{n+1}} + a_{n+1} - a_n$ is an integer for $n \geq N$, i.e., $a_{n+1} \mid a_1 a_n$. By induction, $a_n \mid a_N a_1^{n-N}$ for $n > N$. Let P be the set of all the prime factors of $a_1 a_N$. Then P is a finite set. For $n > N$, the prime factors of a_n are all in P, because $a_n \mid a_1 a_N^{n-N}$.

In order to prove that $\{a_n\}_{n\geq 1}$ is a constant in the end, we only need to prove that $\{v_p(a_n)\}_{n\geq 1}$ is a constant in the end for every prime $p \in P$.

Let $p \in P$ and $n \geq N$. Then one and only one of the following two statements is true:

(1) $v_p(a_{n+1}) \leq v_p(a_n)$;
(2) $v_p(a_{n+1}) > v_p(a_n)$ and $v_p(a_{n+1}) = v_p(a_1)$.

In fact, if $v_p(a_{n+1}) > v_p(a_n)$, then $v_p \left(\frac{a_n}{a_{n+1}} \right) < 0$ and $v_p \left(\frac{a_n}{a_1} \right) < v_p \left(\frac{a_{n+1}}{a_1} \right)$. We know that $v_p \left(\frac{a_n}{a_{n+1}} + \frac{a_{n+1}}{a_1} - \frac{a_n}{a_1} \right) \geq 0$.

Combining it with $v_p \left(\frac{a_n}{a_{n+1}} \right) < 0$, we get that the minimum value in $v_p \left(\frac{a_n}{a_{n+1}} \right), v_p \left(\frac{a_n}{a_1} \right), v_p \left(\frac{a_{n+1}}{a_1} \right)$ appears at least twice, so $v_p \left(\frac{a_n}{a_{n+1}} \right) = v_p \left(\frac{a_n}{a_1} \right)$.

That is, $v_p(a_{n+1}) = v_p(a_1)$. Next, we will prove

(3) If $v_p(a_n) = v_p(a_1)$, where $n > N$ and $p \in P$, then $v_p(a_{n+1}) = v_p(a_1)$.

Assuming $v_p(a_{n+1}) \neq v_p(a_1)$, from (1) and (2) we get $v_p(a_{n+1}) < v_p(a_n) = v_p(a_1)$. Hence,

$$v_p \left(\frac{a_n}{a_{n+1}} \right) > 0, \quad v_p \left(\frac{a_n}{a_1} \right) = 0, \quad v_p \left(\frac{a_{n+1}}{a_1} \right) < 0.$$

Therefore,

$$v_p \left(\frac{a_n}{a_{n+1}} + \frac{a_{n+1}}{a_1} - \frac{a_n}{a_1} \right) = v_p \left(\frac{a_{n+1}}{a_1} \right) < 0,$$

contradicting to the fact that $\frac{a_n}{a_{n+1}} + \frac{a_{n+1}}{a_1} - \frac{a_n}{a_1}$ is an integer.

For any $p \in P$, we have the following two cases:

Case 1: $v_p(a_{n+1}) \leq v_p(a_n)$ for any $n > N$. Then $\{v_p(a_n)\}_{n \geq N}$ is a monotonic non-increasing sequence of non-negative integers, so it will be a constant in the end.

Case 2: There exists $n_0 \geq N$ such that $v_p(a_{n_0+1}) > v_p(a_{n_0})$. From (2) we have $v_p(a_{n_0+1}) > v_p(a_1)$. Then from (3) and by induction, $v_p(a_n) = v_p(a_1)$ for any $n \geq n_0$. Therefore, $\{v_p(a_n)\}$ is a constant in the end.

Remark. For problems concerning divisibility of integers, discussing the power exponent of prime factors is a useful skill, which has appeared many times above. In this proof, we applied the following idea repeatedly: if multiple fractions add up to an integer, then it is impossible that only one term has a unique minimum negative power exponent for some prime factor (i.e., it cannot happen that some prime number only appears in the denominator of one fraction).

【Score Situation】 This particular problem saw the following distribution of scores among contestants: 172 contestants scored 7 points, 11 contestants scored 6 points, 8 contestants scored 5 points, 6 contestants scored 4 points, 7 contestants scored 3 points, 31 contestants scored 2 points, 184 contestants scored 1 point, and 175 contestants scored 0 point. The average score of this problem is 2.695, indicating that it had a certain level of difficulty.

Among the top five teams in the team scores, the scores of this problem are as follows: The United States team scored 42 points (with a total team score of 212 points), the Russia team scored 41 points (with a total team score of 201 points), the China team scored 42 points (with a total team score of 199 points), the Ukraine team scored 42 points (with a total team score of 186 points), and the Thailand team scored 42 points (with a total team score of 183 points).

The gold medal cutoff for this IMO was set at 31 points (with 48 contestants earning gold medals), the silver medal cutoff was 25 points (with 98 contestants earning silver medals), and the bronze medal cutoff was 16 points (with 143 contestants earning bronze medals).

In this IMO, only two contestants achieved a perfect score of 42 points, namely Agnijo Banerjee from the United Kingdom and James Lin from the United States.

Problem 1.23 (IMO 61-5, proposed by Estonia). A deck of $n > 1$ cards is given. A positive integer is written on each card. The deck has the property that the arithmetic mean of the numbers on each pair of cards is also the geometric mean of the numbers on some collection of one or more cards.

For which n does it follow that the numbers on the cards are all equal?

Solution. For all $n > 1$, it is necessary that all cards have the same number.

Equivalently, we prove the following proposition: if $a_1 \leq a_2 \leq \cdots \leq a_n$ are positive integers that are not all equal, then there are two numbers, whose arithmetic mean does not equal the geometric mean of one or more numbers among them.

Let $d = \gcd(a_1, \ldots, a_n)$. If $d > 1$, then we may replace $a_1, a_2, \ldots, a_n$ with $\frac{a_1}{d}, \frac{a_2}{d}, \ldots, \frac{a_n}{d}$. Thus, all the arithmetic and geometric means are divided by d, so the proposition remains equivalent. Hence, we may assume that $a_1, a_2, \ldots, a_n$ are coprime.

Since $a_1, a_2, \ldots, a_n$ are not all equal, $a_n \geq 2$. Let p be a prime factor of a_n. Then there exists some $1 \leq k \leq n - 1$ such that $p \nmid a_k$. Among such k we choose the maximal one. Then we prove that $\frac{a_n + a_k}{2}$ is not the geometric mean of any part of $a_1, a_2, \ldots, a_n$.

Assume for contradiction that there exist $1 \leq i_1 < \cdots < i_m \leq n$ such that $\sqrt[m]{a_{i_1} a_{i_2} \cdots a_{i_m}} = \frac{a_k + a_n}{2}$, so that

$$2^m a_{i_1} a_{i_2} \cdots a_{i_m} = (a_k + a_n)^m.$$

If $i_m > k$, then $p \mid a_{i_m}$, $p \nmid a_k + a_n$, so the left-hand side of the above equality is divisible by p while the right-hand side is not, a contradiction.

If $i_m \leq k$, then since $a_k < a_n$ we have

$$\sqrt[m]{a_{i_1} a_{i_2} \cdots a_{i_m}} \leq a_{i_m} \leq a_k < \frac{a_k + a_n}{2},$$

which is also a contradiction. Therefore, the proposition is proven.

【Score Situation】 This particular problem saw the following distribution of scores among contestants: 223 contestants scored 7 points, 4 contestants scored 6 points, 9 contestants scored 5 points, 1 contestant scored 4 points, 2 contestants scored 3 points, no contestant scored 2 points, 83 contestants scored 1 point, and 294 contestants scored 0 point. The average score of this problem is 2.797, indicating that it had a certain level of difficulty.

Among the top five teams in the team scores, the scores of this problem are as follows: The China team scored 42 points (with a total team score of 215 points), the Russia team scored 36 points (with a total team score of 185 points), the United States team scored 42 points (with a total team score of 183 points), the South Korea team scored 36 points (with a total team score of 175 points), and the Thailand team scored 36 points (with a total team score of 174 points).

The gold medal cutoff for this IMO was set at 31 points (with 49 contestants earning gold medals), the silver medal cutoff was 24 points (with 112 contestants earning silver medals), and the bronze medal cutoff was 15 points (with 155 contestants earning bronze medals).

In this IMO, only one contestant achieved a perfect score of 42 points, namely Jinmin Li from China.

1.2.3 *Function related problems*

Problem 1.24 (IMO 37-3, proposed by Romania). Let S denote the set of non-negative integers. Find all functions f from S to itself such that

$$f(m + f(n)) = f(f(m)) + f(n), \quad \forall m, n \in S.$$

Solution. In the given equation, let $m = n = 0$, and we have

$$f(f(0)) = f(f(0)) + f(0).$$

Thus, $f(0) = 0$. Let $m = 0$. Then

$$f(f(n)) = f(n). \tag{1}$$

Therefore, for any $m, n \in S$ it holds that

$$f(m + f(n)) = f(m) + f(n). \tag{2}$$

Suppose the range of f is $T \subseteq S$. Then by (1), every $f(n)$ in T is a fixed point of f. Evidently, every fixed point of f is in the range of f. Thus, T is actually the set of all fixed points of f.

For any $m \geq n \in T$, it holds that $f(m) = m$ and $f(n) = n$. Then

$$f(m + n) = f(m + f(n)) = f(m) + f(n) = m + n,$$

and hence $m + n \in T$. Moreover,

$$f(m - n) + f(n) = f(m - n + f(n)) = f(m - n + n) = f(m),$$

so $f(m - n) = f(m) - f(n) = m - n$, $m - n \in T$.

If $T \neq \{0\}$, let d be the least positive number in T. For any $c \in T$, suppose $c = kd + r$, where k is a positive integer and $0 \leq r < d$. Then $kd \in T$ and $r = c - kd \in T$. Notice that d is the least positive number in T, and so we have $r = 0$, i.e., every number c in T is a multiple of d.

For any non-negative integer n, let $n = dq + t$, where q is a non-negative integer and $0 \leq t < d$. Then from the above conclusion, $f(t) = h_t d$, where h_t is a non-negative integer. Thus,

$$f(n) = f(dq + t)$$
$$= f((q-1)d + t) + f(d)$$
$$= f((q-1)d + t) + d = \cdots$$
$$= qd + f(t)$$
$$= qd + h_t d,$$

which implies that the functions we want must be in the following form:

$$f(n) = qd + h_t d,$$

where q and t are decided by $n = dq + t$ and $0 \leq t < d$; $h_t(t = 1, 2, \ldots, d-1)$ can be any non-negative integer and $h_0 = 0$.

Finally, we will check if these functions satisfy the given equation. For any $m, n \in S$, suppose $m = dq + t$ and $f(n) = kd$. Then

$$f(m + f(n)) = f(dq + t + kd)$$
$$= f((q+k)d + t)$$
$$= (q+k)d + f(t)$$
$$= qd + f(t) + kd$$
$$= f(m) + f(n).$$

Remark. The key to this problem is to find the fixed point and the related properties. The subsequent part seems complicated, but in fact it is an imitation of the Euclidean algorithm, which is a basic method in number theory problems.

【Score Situation】This particular problem saw the following distribution of scores among contestants: 52 contestants scored 7 points, 19 contestants scored 6 points, 19 contestants scored 5 points, 14 contestants scored 4 points, 22 contestants scored 3 points, 73 contestants scored 2 points, 176 contestants scored 1 point, and 49 contestants scored 0 point. The average score of this problem is 2.399, indicating that it had a certain level of difficulty.

Among the top five teams in the team scores, the scores of this problem are as follows: The Romania team scored 37 points (with a total team score of 187 points), the United States team scored 36 points (with a total team score of 185 points), the Hungary team scored 34 points (with a total team score of 167 points), the Russia team scored 33 points (with a total team score of 162 points), and the United Kingdom team scored 35 points (with a total team score of 161 points).

The gold medal cutoff for this IMO was set at 28 points (with 35 contestants earning gold medals), the silver medal cutoff was 20 points (with 66 contestants earning silver medals), and the bronze medal cutoff was 12 points (with 99 contestants earning bronze medals).

In this IMO, only one contestant achieved a perfect score of 42 points, namely Ciprian Manolescu from Romania.

Problem 1.25 (IMO 38-6, proposed by Lithuania). For each positive integer n, let $f(n)$ denote the number of ways of representing n as a sum of powers of 2 with non-negative integer exponents. Representations which differ only in the ordering of their summands are considered to be the same. For instance, $f(4) = 4$, because the number 4 can be represented in the following four ways: $4; 2 + 2; 2 + 1 + 1; 1 + 1 + 1 + 1$.

Prove that, for any integer $n \geq 3, 2^{\frac{n^2}{4}} < f(2^n) < 2^{\frac{n^2}{2}}$.

Proof. First we consider the recurrence relations of $f(n)$.

For any positive odd integer $n = 2k + 1$ greater than 1, every representation of n must contain at least one 1, which can be removed and then a representation of $2k$ is obtained; conversely, we can adjoin one 1 to a representation of $2k$ to get a representation of $2k + 1$. There exists a one-to-one correspondence between the representations of $2k + 1$ and $2k$. Thus

$$f(2k + 1) = f(2k). \tag{1}$$

For any positive even integer $n = 2k$, the representations of n can be divided into two cases: if a representation contains at least one 1, then by removing it we obtain a representation of $2k - 1$; if there is no 1 in the representation, by dividing all terms by 2 we get a representation of k. These two kinds of operations are both mutually inverses, so we find two disjoint one-to-one correspondences. Therefore,

$$f(2k) = f(2k - 1) + f(k). \tag{2}$$

From (1) and (2), function f is nondecreasing. For positive integers $k = 2, 3, \ldots$,

$$f(2k) - f(2k - 2) = f(2k) - f(2k - 1) = f(k).$$

For any positive integer $n \geq 2$, we can add up the above equalities for $k = 2, 3, \ldots$, and obtain

$$f(2n) = f(2) + f(2) + f(3) + \cdots + f(n). \tag{3}$$

Now, we look into the upper estimate. Evidently, $f(2) = 2$. From (3), we have $f(2n) \leq n \cdot f(n)$, so for positive integers $n \geq 3$,

$$f(2^n) \leq 2^{n-1} \cdot f(2^{n-1})$$

$$\leq 2^{n-1} \cdot 2^{n-2} \cdot f(2^{n-2}) \leq \cdots$$

$$\leq 2^{(n-1)+(n-2)+\cdots+1} \cdot f(2)$$

$$= 2^{\frac{n(n-1)}{2}} \cdot 2$$

$$= 2^{\frac{n^2}{2}+\frac{2-n}{2}} < 2^{\frac{n^2}{2}}.$$

Finally, we turn the perspective to the lower estimate. To obtain the lower estimate we will prove that for any positive integers $n \geq 2$,

$$f(1) + f(2) + \cdots + f(2n) \geq 2n \cdot f(n). \tag{4}$$

For positive integers $b \geq a \geq 1$ with the same parity,

$$f(b+1) - f(b) \geq f(a+1) - f(a). \tag{5}$$

(If both a and b are even, then by (1), both sides of (5) are equal to 0; if both a and b are odd, then by (2), we have $f(b+1) - f(b) = f\left(\frac{b+1}{2}\right) \geq f\left(\frac{a+1}{2}\right) = f(a+1) - f(a)$.) Then for $m = 1, 2, \ldots, n-1$, since $n+m$ and $n-m$ have the same parity, from (5) we obtain

$$f(n+m+1) - f(n+m) \geq f(n-m+1) - f(n-m),$$

i.e.,

$$f(n+m+1) + f(n-m) \geq f(n+m) + f(n-m+1).$$

Hence,

$$f(2n) + f(1) \geq f(2n-1) + f(2) \geq \cdots \geq f(n+1) + f(n) \geq 2f(n).$$

Adding up these inequalities proves (4).

By (3), the left-hand side of (4) is actually $f(4n) - 1$, so

$$f(4n) > 2n \cdot f(n). \tag{6}$$

We can see $f(2^2) = 4 > 2^{\frac{2^2}{4}}$ and $f(2^3) = f(2) + f(2) + f(3) + f(4) = 10 > 2^{\frac{3^2}{4}}$. When $n \geq 4$, assume $f(2^{n-2}) > 2^{\frac{(n-2)^2}{4}}$ for $n - 2$. Then using (6) we have

$$f(2^n) > 2^{n-1} \cdot f(2^{n-2}) > 2^{n-1+\frac{(n-2)^2}{4}} = 2^{\frac{n^2}{4}}.$$

By mathematical induction we obtain the lower estimate as required.

Summarizing, for $n \geq 3$, we see that $2^{\frac{n^2}{4}} < f(2^n) < 2^{\frac{n^2}{2}}$.

Remark. For those functions or sequences whose expression is difficult to be given, we may try to study a recurrence formula instead.

【Score Situation】 This particular problem saw the following distribution of scores among contestants: 10 contestants scored 7 points, 4 contestants scored 6 points, 27 contestants scored 5 points, 4 contestants scored 4 points, 22 contestants scored 3 points, 12 contestants scored 2 points, 40 contestants scored 1 point, and 341 contestants scored 0 point. The average score of this problem is 0.815, indicating that it was extremely difficult.

Among the top five teams in the team scores, the scores of this problem are as follows: The China team scored 25 points (with a total team score of 223 points), the Hungary team scored 25 points (with a total team score of 219 points), the Iran team scored 16 points (with a total team score of 217 points), the Russia team scored 21 points (with a total team score of 202 points), and the United States team scored 16 points (with a total team score of 202 points).

The gold medal cutoff for this IMO was set at 35 points (with 39 contestants earning gold medals), the silver medal cutoff was 25 points (with 70 contestants earning silver medals), and the bronze medal cutoff was 15 points (with 122 contestants earning bronze medals).

In this IMO, a total of four contestants achieved a perfect score of 42 points.

Problem 1.26 (IMO 39-6, proposed by Bulgaria). Consider all functions f from the set N of all positive integers into itself satisfying $f(t^2 f(s)) = s(f(t))^2$ for all s and t in N. Determine the least possible value of $f(1998)$.

Solution. First, we prove that the given equation $f(t^2 f(s)) = s(f(t))^2$ is satisfied if and only if the following two equations hold simultaneously:

$$f(f(s)) = s(f(1))^2, \tag{1}$$

$$f(st) = \frac{f(s)f(t)}{f(1)}. \tag{2}$$

By (1) and (2), we have

$$f(t^2 f(s)) = \frac{f(t)f(t \cdot f(s))}{f(1)}$$

$$= \frac{f(t) \cdot f(t)f(f(s))}{(f(1))^2}$$

$$= s(f(t))^2,$$

which is the given condition. So these two equations are sufficient.

Moreover, if the given equation holds, by letting $t = 1$, then $f(f(s)) = s(f(1))^2$, namely (1) holds. Substituting $f(s)$ for s, we have

$$f(t^2 s(f(1))^2) = f(t^2 f(f(s))) = f(s)(f(t))^2. \tag{3}$$

Using equation (3), we first set $s = t^2, t = 1$ and then set $s = 1$, to obtain

$$f(t^2)(f(1))^2 = f(t^2(f(1))^2) = f(1)(f(t))^2,$$

and hence $f(t^2) = \frac{(f(t))^2}{f(1)}$, so that

$$f(s^2 t^2) = \frac{(f(st))^2}{f(1)}. \tag{4}$$

By equations (3) and (4),

$$f(s^2 t^2)(f(1))^2 = f(1)(f(st))^2$$

$$= f(s^2 t^2 (f(1))^2)$$

$$= f(t^2)(f(s))^2$$

$$= \frac{(f(t))^2}{f(1)} \cdot (f(s))^2.$$

Together with (4) we can see (2) holds. So, equations (1) and (2) are necessary.

Next, we prove that: for any $t \in \mathbf{N}^*$, it holds that $f(1) \mid f(t)$.

Set $s = t$ in equation (2). By induction, $f(t^n) = \left(\frac{f(t)}{f(1)}\right)^{n-1} f(t), n \in \mathbf{N}^*$. Assume for contradiction that $f(1) \nmid f(t)$. Then there exists a prime factor p of $f(1)$ and a positive integer m, such that $p^m \mid f(1)$ but $p^m \nmid f(t)$. When $n \geq m + 1$, there is $p^{m(n-1)} \nmid (f(t))^n$, and then $f(t^n) \notin \mathbf{N}^*$, a contradiction.

Hence, we can suppose $f(n) = k_n \cdot f(1)$ with $k_n \in \mathbf{N}^*$. If $k_s = k_t$, then

$$s(f(1))^2 = f(f(s)) = f(f(t)) = t(f(1))^2,$$

so $s = t$. This is to say, for $s \neq t$, we have $k_s \neq k_t$. In particular, $k_n > 1$ when $n > 1$.

Now, we work on finding the least possible value of $f(1998)$. First,

$$f(1998) = f(2 \times 3^3 \times 37)$$

$$= \frac{f(2)(f(3))^3 f(37)}{(f(1))^4}$$

$$= k_2 \cdot k_3^3 \cdot k_{37} f(1),$$

where k_2, k_3, k_{37} are greater than 1 and different from each other. Therefore,

$$f(1998) \geq 1 \times 2^3 \times 3 \times 5 = 120.$$

Here we only need to mention that $f(1998) \neq 1 \times 2^3 \times 3 \times 4$. Otherwise, we must have $f(3) = 2$ and $f(1) = 1$. Then $f(9) = \frac{(f(3))^2}{f(1)} = 4$, so both k_2 and k_{37} cannot be 4, a contradiction.

Finally, we construct the function f such that $f(1998) = 120$. Set $f(1) = 1$, $f(2) = 3$, $f(3) = 2$, $f(5) = 37$, $f(37) = 5$, and $f(p) = p$ (for all the prime numbers p different from 2, 3, 5, 37). For every composite number $n = p_1^{\alpha_1} p_2^{\alpha_2} \cdots p_r^{\alpha_r}$, let

$$f(n) = f(p_1)^{\alpha_1} f(p_2)^{\alpha_2} \cdots f(p_r)^{\alpha_r}.$$

We can easily verify that this function f satisfies both (1) and (2), and then it also satisfies the original given equation. We conclude that the least possible value of $f(1998)$ is 120.

Remark. This is a difficult problem of functional equations, because from the given condition we cannot obtain helpful properties in a simple way. Nevertheless, our approach is still similar to solving other functional equation problems, which is to obtain new information by repeatedly substituting and iterating. Combining with the divisibility properties in number theory we usually can simplify the condition to a processable form (here it is $f(n) = k_n \cdot f(1)$). It is worth mentioning that the function solutions of this problem actually need and only need to determine all the function values for prime numbers. This is consistent with the important idea in number theory that all positive integers are derived from prime numbers.

【Score Situation】This particular problem saw the following distribution of scores among contestants: 24 contestants scored 7 points, 2 contestants scored 6 points, 1 contestant scored 5 points, 7 contestants scored 4 points, 10 contestants scored 3 points, 6 contestants scored 2 points, 29 contestants scored 1 point, and 340 contestants scored 0 point. The average score of this problem is 0.678, indicating that it was extremely difficult.

Among the top five teams in the team scores, the scores of this problem are as follows: The Iran team scored 35 points (with a total team score of 211 points), the Bulgaria team scored 36 points (with a total team score of 195 points), the Hungary team scored 10 points (with a total team score of 186 points), the United States team scored 25 points (with a total team score of 186 points), and the Chinese Taiwan team scored 19 points (with a total team score of 184 points).

The gold medal cutoff for this IMO was set at 31 points (with 37 contestants earning gold medals), the silver medal cutoff was 24 points (with 66 contestants earning silver medals), and the bronze medal cutoff was 14 points (with 102 contestants earning bronze medals).

In this IMO, only one contestant achieved a perfect score of 42 points, namely Omid Amini from Iran.

Problem 1.27 (IMO 51-3, proposed by the United States). Let $\mathbf{N}^*$ be the set of positive integers. Determine all functions $g : \mathbf{N}^* \to \mathbf{N}^*$ such that

$$(g(m) + n)(m + g(n))$$

is a perfect square for all $m, n \in \mathbf{N}^*$.

Solution. The answer is $g(n) = n + C$, where C is a non-negative integer. Clearly, the function $g(n) = n + C$ satisfies the required property since

$$(g(m) + n)(m + g(n)) = (n + m + C)^2$$

is a perfect square. We first prove a lemma.

Lemma. *If a prime number p divides $g(k) - g(l)$ for some positive integers k and l, then $p \mid k - l$.*

Proof of the lemma. If $p^2 \mid g(k) - g(l)$, let $g(l) = g(k) + p^2 a$, where a is an integer. Choose an integer $D > \max\{g(k), g(l)\}$, and D is not divisible by p. Set $n = pD - g(k)$; then $n + g(k) = pD$, and thus

$$n + g(l) = pD + (g(l) - g(k)) = p(D + pa)$$

is divisible by p, but not divisible by p^2.

By assumption, $(g(k) + n)(g(n) + k)$ and $(g(l) + n)(g(n) + l)$ are both perfect squares, and therefore they are divisible by p^2 since they are divisible by p. Hence, $p \mid ((g(n) + k)), p \mid ((g(n) + l))$. Therefore,

$$p \mid ((g(n) + k) - (g(n) + l)),$$

i.e., $p \mid k - l$.

If $p \mid g(k) - g(l)$ but p^2 does not divide $g(k) - g(l)$, choose an integer D as above and set $n = p^3 D - g(k)$. Then $g(k) + n = p^3 D$ is divisible by p^3, but not by p^4, and $g(l) + n = p^3 D + (g(l) - g(k))$ is divisible by p, but not by p^2. As with the above argument, we have $p \mid g(n) + k$ and $p \mid g(n) + l$. Therefore,

$$p \mid ((g(n) + k) - (g(n) + l)),$$

i.e., $p \mid k - l$. This completes the proof of the lemma.

Back to the original problem: if there exist positive integers k and l such that $g(k) = g(l)$, then the lemma implies that $k - l$ is divisible by any prime number. Hence, $k - l = 0$, i.e., $k = l$, and thus g is injective.

Now consider $g(k)$ and $g(k + 1)$. Since $(k + 1) - k = 1$, once again the lemma implies that $g(k + 1) - g(k)$ is not divisible by any prime number, and therefore $\mid g(k + 1) - g(k) \mid = 1$.

Let $g(2) - g(1) = q$, where $|q| = 1$. It follows easily by induction that $g(n) = g(1) + (n - 1)q$.

If $q = -1$, then $g(n) \le 0$ for $n \ge g(1) + 1$, a contradiction. Therefore, we must have $q = 1$ and $g(n) = n + (g(1) - 1)$ for any $n \in \mathbf{N}^*$, where $g(1) - 1 \ge 0$. Set $g(1) - 1 = C$ (a constant). Then $g(n) = n + C$, where C is a non-negative integer.

【Score Situation】 This particular problem saw the following distribution of scores among contestants: 16 contestants scored 7 points, 2 contestants scored 6 points, 4 contestants scored 5 points, 4 contestants scored 4 points, 4 contestants scored 3 points, 10 contestants scored 2 points, 48 contestants scored 1 point, and 428 contestants scored 0 point. The average score of this problem is 0.465, indicating that it was extremely difficult.

Among the top five teams in the team scores, the scores of this problem are as follows: The China team scored 23 points (with a total team score of 197 points), the Russia team scored 11 points (with a total team score of 169 points), the United States team scored 8 points (with a total team score of 168 points), the South Korea team scored 24 points (with a total team score of 156 points), the Thailand team scored 16 points (with a total team score of 148 points), and the Kazakhstan team scored 13 points (with a total team score of 148 points).

The gold medal cutoff for this IMO was set at 27 points (with 47 contestants earning gold medals), the silver medal cutoff was 21 points (with 103 contestants earning silver medals), and the bronze medal cutoff was 15 points (with 115 contestants earning bronze medals).

In this IMO, only one contestant achieved a perfect score of 42 points, namely Zipei Nie from China.

Problem 1.28 (IMO 52-5, proposed by Iran). Let f be a function from the set of integers to the set of positive integers. Suppose that, for any two integers m and n, the difference $f(m) - f(n)$ is divisible by $f(m-n)$. Prove that, for all integers m and n with $f(m) \leq f(n)$, the number $f(n)$ is divisible by $f(m)$.

Proof. According to the assumption, $f(k) \mid (f(x+k) - f(x)), x \in \mathbf{Z}$ for any integer $k \neq 0$. Consequently, $f(k) \mid (f(x+tk) - f(x)), \forall x \in \mathbf{Z}, t \in \mathbf{Z}$. So $f(k) \mid (f(m) - f(n))$ for any m and n such that $k \mid (m-n)$.

We now show that $f(n) = f(-n)$ for any integer n. Since otherwise, without loss of generality, if there is an integer $n \neq 0$ such that $f(n) > f(-n) > 0$, then $0 < f(n) - f(-n) < f(n)$, which contradicts $f(n) \mid (f(n) - f(-n))$. Thus, $f(m - (-n)) \mid (f(m) - f(-n))$. Therefore, for $f(m) < f(n)$,

$$f(n) \mid f(m+n) - f(m), \tag{1}$$

$$f(m+n) \mid f(m) - f(n). \tag{2}$$

If $0 < f(m) < f(n)$, then by (2), we have $0 < f(m+n) < f(n)$. Combining it and (1), we obtain $f(m+n) = f(m)$. Then by (2), we get the result $f(m) \mid f(n)$. For $f(m) = f(n)$, the result is obvious.

Remark. The given condition may seem similar to the proof and use of the lemma in the previous problem. However, they differ in many details. In functional equation problems, we can start with the basic properties of functions. For example, in this problem, we needed to check whether the desired function is even. Afterwards, both the proof and the use of this conclusion are still difficult, but the problem has been simplified a lot.

【Score Situation】 This particular problem saw the following distribution of scores among contestants: 170 contestants scored 7 points, 19 contestants scored 6 points, 9 contestants scored 5 points, 20 contestants scored 4 points, 20 contestants scored 3 points, 127 contestants scored 2 points, 92 contestants scored 1 point, and 106 contestants scored 0 point. The average score of this problem is 3.259, indicating that it was relatively straightforward.

Among the top five teams in the team scores, the scores of this problem are as follows: The China team scored 42 points (with a total team score of 189 points), the United States team scored 42 points (with a total team score of 184 points), the Singapore team scored 36 points (with a total team score of 179 points), the Russia team scored 42 points (with a total team score of 161 points), and the Thailand team scored 42 points (with a total team score of 160 points).

The gold medal cutoff for this IMO was set at 28 points (with 54 contestants earning gold medals), the silver medal cutoff was 22 points (with 90 contestants earning silver medals), and the bronze medal cutoff was 16 points (with 137 contestants earning bronze medals).

In this IMO, only one contestant achieved a perfect score of 42 points, namely Lisa Sauermann from Germany.

1.2.4 *Other problems*

Problem 1.29 (IMO 10-2, proposed by Czechoslovakia). Find all natural numbers x such that the product of their digits (in the decimal notation) is equal to $x^2 - 10x - 22$.

Solution. Suppose the decimal representation of x is

$$x = a_n \cdot 10^n + a_{n-1} \cdot 10^{n-1} + \cdots + a_1 \cdot 10 + a_0,$$

where $a_0, a_1, \ldots, a_n \in \{0, 1, 2, \ldots, 9\}$, and $a_n \neq 0$.

Since $x^2 - 10x - 22 \geq 0$, then $x > 11$. On the other hand,

$$x^2 - 10x - 22 = a_n \cdot a_{n-1} \cdots \cdot a_1 \cdot a_0 \leq a_n \cdot 9^n$$

$$< a_n \cdot 10^n + a_{n-1} \cdot 10^{n-1} + \cdots + a_1 \cdot 10 + a_0 = x,$$

i.e., $x^2 - 11x - 22 < 0$,

so

$$x < \frac{11 + \sqrt{209}}{2} < 13.$$

Hence, $x = 12$.

When $x = 12$, the number $x^2 - 10x - 22 = 1 \cdot 2$ satisfies the given condition. Conclusively, the solution is $x = 12$.

【Score Situation】This particular problem saw the following distribution of scores among contestants: 59 contestants scored 7 points, 14 contestants scored 6 points, 6 contestants scored 5 points, 2 contestants scored 4 points, 4 contestants scored 3 points, 1 contestant scored 2 points, 3 contestants scored 1 point, and 2 contestants scored 0 point. The average score of this problem is 6.066, indicating that it was simple.

Among the top five teams in the team scores, the German Democratic Republic team achieved a total score of 304 points, the Soviet Union team achieved a total score of 298 points, the Hungary team achieved a total score of 291 points, the United Kingdom team achieved a total score of 263 points, and the Poland team achieved a total score of 262 points.

The gold medal cutoff for this IMO was set at 39 points (with 22 contestants earning gold medals), the silver medal cutoff was 33 points (with 22 contestants earning silver medals), and the bronze medal cutoff was 26 points (with 20 contestants earning bronze medals).

In this IMO, a total of 16 contestants achieved a perfect score of 40 points.

Problem 1.30 (IMO 10-6, proposed by the United Kingdom). For every natural number n, evaluate the sum

$$\sum_{k=0}^{\infty} \left[\frac{n+2^k}{2^{k+1}} \right] = \left[\frac{n+1}{2} \right] + \left[\frac{n+2}{4} \right] + \cdots + \left[\frac{n+2^k}{2^{k+1}} \right] + \cdots.$$

(The symbol $[x]$ denotes the greatest integer not exceeding x.)

Solution 1. For any positive integer n, there exists one and only one nonnegative integer r such that $2^r \leq n < 2^{r+1}$. For this r,

$$\sum_{k=0}^{\infty} \left[\frac{n+2^k}{2^{k+1}} \right] = \sum_{k=0}^{r} \left[\frac{n+2^k}{2^{k+1}} \right].$$

Let $a_k = \left[\frac{n+2^k}{2^{k+1}} \right]$ $(k = 0, 1, \ldots, r)$. Then a_k is the largest positive integer m such that $m \leq \frac{n+2^k}{2^{k+1}}$, i.e., $2^k(2m-1) \leq n$. Now we take a look at sequence $\{2^k(2m-1)\}(k = 0, 1, \ldots, r, m = 1, 2, \ldots, a_k)$. By the definition of the a_k, the number of terms is $\sum_{k=0}^{r} a_k$.

On the other hand, for any positive integer less than or equal to n, there exists one and only one way to represent it in the form $2^k(2m-1)$, where $k = 0, 1, \ldots, r$ and $m = 1, 2, \ldots, a_k$. That is to say, every number in $1, 2, \ldots, n$ appears exactly once in the sequence $\{2^k(2m-1)\}$. Hence the number of terms of this sequence is n.

Summarizing, we have $\sum_{k=0}^{r} a_k = n$.

Solution 2. Suppose the binary representation of n is

$$n = a_0 \cdot 2^r + a_1 \cdot 2^{r-1} + \cdots + a_{r-1} \cdot 2 + a_r,$$

where $a_0 = 1$ and $a_i = 0, 1$ $(i = 1, \ldots, r)$.

When $k = 0, 1, \ldots, r$, we have $\left[\frac{n+2^k}{2^{k+1}}\right] = a_0 \cdot 2^{r-k-1} + a_1 \cdot 2^{r-k-2} + \cdots$ $+ a_{r-k-1} \cdot 2^0 + a_{r-k}$; when $k > r$, we see that $\left[\frac{n+2^k}{2^{k+1}}\right] = 0$. Therefore,

$$\sum_{k=0}^{\infty} \left[\frac{n+2^k}{2^{k+1}}\right] = \sum_{k=0}^{r} \left[\frac{n+2^k}{2^{k+1}}\right]$$

$$= \sum_{k=0}^{r} \left(a_0 \cdot 2^{r-k-1} + a_1 \cdot 2^{r-k-2} + \cdots + a_{r-k-1} \cdot 2^0 + a_{r-k}\right)$$

$$= a_0 \cdot (1 + 1 + 2 + \cdots + 2^{r-1}) + a_1 \cdot (1 + 1 + 2 + \cdots + 2^{r-2})$$

$$+ \cdots + a_{r-1} \cdot (1+1) + a_r$$

$$= a_0 \cdot 2^r + a_1 \cdot 2^{r-1} + \cdots + a_{r-1} \cdot 2 + a_r = n.$$

Solution 3. First we prove a lemma.

Lemma. *For any real number x,*

$$\left[x + \frac{1}{2}\right] = [2x] - [x].$$

Proof of the Lemma. Let $[x] = m$.

If $m \le x < m + \frac{1}{2}$, then $\left[x + \frac{1}{2}\right] = m = 2m - m = [2x] - [x]$.
If $m + \frac{1}{2} \le x < m + 1$, then $\left[x + \frac{1}{2}\right] = m + 1 = 2m + 1 - m = [2x] - [x]$.
The lemma is proved.

Back to the original problem, for any positive integer n, there exists one and only one non-negative integer r such that $2^r \le n < 2^{r+1}$. Hence

$$\sum_{k=0}^{\infty} \left[\frac{n+2^k}{2^{k+1}}\right] = \sum_{k=0}^{r} \left[\frac{n+2^k}{2^{k+1}}\right]$$

$$= \sum_{k=0}^{r} \left[\frac{n}{2^{k+1}} + \frac{1}{2}\right]$$

$$= \sum_{k=0}^{r} \left(\left[\frac{n}{2^k}\right] - \left[\frac{n}{2^{k+1}}\right]\right) \quad \text{(By the lemma)}$$

$$= \left[\frac{n}{2^0}\right] - \left[\frac{n}{2^{r+1}}\right] = n.$$

Remark. There were three completely different methods to solve this problem. In the first method, we used the definition of floor functions and a common expression of integers to construct a one-to-one correspondence. The second method requires familiarity with the binary representation, while the third method (provided by a contestant of that IMO) requires familiarity with the properties of floor functions.

【Score Situation】This particular problem saw the following distribution of scores among contestants: 43 contestants scored 8 points, 10 contestants scored 7 points, 1 contestant scored 6 points, 5 contestants scored 5 points, no contestant scored 4 points, 7 contestants scored 3 points, 3 contestants scored 2 points, 12 contestants scored 1 point, and 10 contestants scored 0 point. The average score of this problem is 5.319, indicating that it was simple.

Among the top five teams in the team scores, the German Democratic Republic team achieved a total score of 304 points, the Soviet Union team achieved a total score of 298 points, the Hungary team achieved a total score of 291 points, the United Kingdom team achieved a total score of 263 points, and the Poland team achieved a total score of 262 points.

The gold medal cutoff for this IMO was set at 39 points (with 22 contestants earning gold medals), the silver medal cutoff was 33 points (with 22 contestants earning silver medals), and the bronze medal cutoff was 26 points (with 20 contestants earning bronze medals).

In this IMO, a total of 16 contestants achieved a perfect score of 40 points.

Problem 1.31 (IMO 12-2, proposed by Romania). Let a, b, and n be integers greater than 1, and let a and b be the bases of two number systems. A_{n-1} and A_n are numbers in the system with base a, and B_{n-1} and B_n are numbers in the system with base b; these are related as follows:

$$A_{n-1} = x_{n-1}x_{n-2}\cdots x_0, A_n = x_n x_{n-1}\cdots x_0,$$

$$B_{n-1} = x_{n-1}x_{n-2}\cdots x_0, B_n = x_n x_{n-1}\cdots x_0,$$

where $x_n \neq 0, x_{n-1} \neq 0$.

Prove: $\frac{A_{n-1}}{A_n} < \frac{B_{n-1}}{B_n}$ if and only if $a > b$.

Proof. Since $A_n = x_n a^n + A_{n-1}$ and $B_n = x_n b^n + B_{n-1}$, then

$$\frac{A_{n-1}}{A_n} < \frac{B_{n-1}}{B_n} \Leftrightarrow \frac{A_n}{A_{n-1}} > \frac{B_n}{B_{n-1}}$$

$$\Leftrightarrow \frac{x_n a^n + A_{n-1}}{A_{n-1}} > \frac{x_n b^n + B_{n-1}}{B_{n-1}}$$

$$\Leftrightarrow \frac{x_n a^n}{A_{n-1}} > \frac{x_n b^n}{B_{n-1}}$$

$$\Leftrightarrow a^n B_{n-1} - b^n A_{n-1} > 0$$

$$\Leftrightarrow a^n \cdot \sum_{k=0}^{n-1} x_k b^k - b^n \cdot \sum_{k=0}^{n-1} x_k a^k > 0$$

$$\Leftrightarrow \sum_{k=0}^{n-1} x_k a^k b^k (a^{n-k} - b^{n-k}) > 0$$

$$\Leftrightarrow a > b.$$

The last step is because that for any $k = 0, 1, \ldots, n-1$, when $a, b > 1$ there is the equivalence $a^{n-k} > b^{n-k} \Leftrightarrow a > b$.

【Score Situation】 This particular problem saw the following distribution of scores among contestants: 21 contestants scored 7 points, 5 contestants scored 6 points, 2 contestants scored 5 points, 3 contestants scored 4 points, 3 contestants scored 3 points, no contestant scored 2 points, no contestant scored 1 point, and 1 contestant scored 0 point. The average score of this problem is 5.943, indicating that it was simple.

Among the top five teams in the team scores, the Hungary team achieved a total score of 233 points, the German Democratic Republic team achieved a total score of 221 points, the Soviet Union team achieved a total score of 221 points, the Yugoslavia team achieved a total score of 209 points, and the Romania team achieved a total score of 208 points.

The gold medal cutoff for this IMO was set at 37 points (with 7 contestants earning gold medals), the silver medal cutoff was 30 points (with 11 contestants earning silver medals), and the bronze medal cutoff was 19 points (with 40 contestants earning bronze medals).

In this IMO, only three contestants achieved a perfect score of 40 points, namely Wolfgang Burmeister from the German Democratic Republic, Imre Ruzsa from Hungary and Andrei Hodulev from the Soviet Union.

Problem 1.32 (IMO 33-6, proposed by the United Kingdom). For each positive integer n, the number $S(n)$ is defined to be the greatest integer such that n^2 can be written as the sum of k positive squares for every positive integer $k \leq S(n)$.

(a) Prove that $S(n) \leq n^2 - 14$ for each $n \geq 4$.
(b) Find an integer n such that $S(n) = n^2 - 14$.
(c) Prove that there are infinitely many integers n such that $S(n) = n^2 - 14$.

Solution. (a) Assume for contradiction that $S(n) > n^2 - 14$ for some integer $n \geq 4$. Then there exist $k = n^2 - 13$ positive integers $a_1, a_2, \ldots, a_k$, such that

$$n^2 = a_1^2 + a_2^2 + \cdots + a_k^2,$$

i.e., $\sum_i^k (a_i^2 - 1) = 13$. Thus $0 \leq a_i^2 - 1 \leq 13$, so $a_i^2 - 1 = 0, 3, 8$ $(i = 1, 2, \ldots, k)$.

Assume that the numbers of 0, 3, and 8 appearing in the values of $a_i^2 - 1$ are a, b, and c, respectively. Then $3b + 8c = 13$ and $c = 0$ or 1. However, in neither case is b an integer, a contradiction.

(b) We first prove that every positive integer m greater than 13 can be represented in the form of $3b + 8c$, where b and c are non-negative integers. We can easily reach the conclusion when m is divisible by 3. Otherwise, if $m = 3s + 1$, then $s \geq 5$ and $m = 3(s - 5) + 8 \times 2$; if $m = 3s + 2$, then $s \geq 4$ and $m = 3(s - 2) + 8 \times 1$.

Therefore, if m satisfies that $14 \leq m \leq \frac{3}{4}n^2$, then for the b and c defined above,

$$m + b + c \leq m + \frac{1}{3}m \leq n^2,$$

and together with

$$n^2 = (n^2 - m - b - c) \cdot 1^2 + b \cdot 2^2 + c \cdot 3^2$$

we know n^2 can be written as the sum of $n^2 - m$ positive squares, i.e., n^2 can be written as the sum of $n^2 - 14, n^2 - 15, \ldots, \left[\frac{1}{4}n^2\right] + 1$ positive squares.

When $n = 13$, we have $n^2 = 12^2 + 5^2 = 12^2 + 4^2 + 3^2 = 8^2 + 8^2 + 5^2 + 4^2$. Since 8^2 can be written as the sum of four 4^2s, the number 4^2 can be written as the sum of four 2^2s, the number 2^2 can be written as the sum of four 1^2s, so that $13^2 = 8^2 + 8^2 + 5^2 + 4^2$ can be written as the sum of $4, 7, 10, \ldots, 43$ positive squares. Observe that $5^2 = 4^2 + 3^2$, so 13^2 can also be written as the sum of $5, 8, 11, \ldots, 44$ positive squares. Now that 12^2 can be written as the sum of four 6^2s, the number 6^2 can be written as the sum of four 3^2s. Then $13^2 = 12^2 + 4^2 + 3^2$ can be written as the sum of $3, 6, 9, \ldots, 33$ positive squares. Moreover, $13^2 = 17 \times 3^2 + 4^2$ and $3^2 = 2^2 + 2^2 + 1$, so 13^2 can be written as the sum of $18 + 2 \times 9 = 36$ or $18 + 2 \times 12 = 42$ positive squares. Finally, 4^2 can be written as the sum of four 2^2s, thus 13^2 can be written as the sum of 39 positive squares.

Summarizing, we see that 13^2 can be written as the sum of $1, 2, \ldots, 44$ positive squares, and together with $\left[\frac{1}{4} \cdot 13^2\right] = 42$ we know $n = 13$ satisfies $S(n) = n^2 - 14$.

(c) Let $n = 2^k \times 13$, we prove that such n satisfies $S(n) = n^2 - 14$.

First we observe that, if some n^2 can be written as the sum of $1, 2, \ldots, s$ positive squares, then $(2n)^2 = n^2 + n^2 + n^2 + n^2$ can be written as the sum of $1, 2, \ldots, 4s$ positive squares.

Then by the conclusion of (b), $13^2 \times 2^2$ can be written as the sum of $1, 2, \ldots, 155 \times 4$ positive squares, $13^2 \times 2^4$ can be written as the sum of $1, 2, \ldots, 155 \times 4^2$ positive squares, $\cdots$, $n^2 = 13^2 \times 2^{2k}$ can be written as the sum of $1, 2, \ldots, 155 \times 4^k$ positive squares.

At the same time, we have proved that n^2 can be written as the sum of $n^2 - 14, n^2 - 15, \ldots, \left[\frac{1}{4}n^2\right] + 1$ positive integers. Since $155 \times 4^k > \frac{1}{4}(2^k \times 13)^2 = \frac{1}{4}n^2$, then n^2 can be written as the sum of $1, 2, \ldots, n^2 - 14$ positive squares, which implies that $S(n) = n^2 - 14$.

【Score Situation】This particular problem saw the following distribution of scores among contestants: 30 contestants scored 7 points, 8 contestants scored 6 points, 19 contestants scored 5 points, 37 contestants scored 4 points, 40 contestants scored 3 points, 75 contestants scored 2 points, 18 contestants scored 1 point, and 123 contestants scored 0 point. The average score of this problem is 2.254, indicating that it had a certain level of difficulty.

Among the top five teams in the team scores, the scores of this problem are as follows: The China team scored 42 points (with a total team score of 240 points), the United States team scored 33 points (with a total team score of 181 points), the Romania team scored 21 points (with a total team score of 177 points), the Commonwealth of Independent States team scored 29 points (with a total team score of 176 points), and the United Kingdom team scored 32 points (with a total team score of 168 points).

The gold medal cutoff for this IMO was set at 32 points (with 26 contestants earning gold medals), the silver medal cutoff was 24 points (with 55 contestants earning silver medals), and the bronze medal cutoff was 14 points (with 74 contestants earning bronze medals).

In this IMO, a total of four contestants achieved a perfect score of 42 points.

Problem 1.33 (IMO 35-3, proposed by Romania). For any positive integer k, let $f(k)$ be the number of elements in the set $\{k + 1, k + 2, \ldots, 2k\}$ whose base 2 representation has precisely three 1s.

(a) Prove that, for each positive integer m, there exists at least one positive integer k such that $f(k) = m$.

(b) Determine all positive integers m for which there exists exactly one k with $f(k) = m$.

Solution. (a) Denote by S the set of those positive integers whose base 2 representation has precisely three 1s. Then for any positive integer k, we see that $f(k)$ is the number of elements of $\{k+1, k+2, \ldots, 2k\} \cap S$ and $f(k+1)$ is the number of elements of $\{k+2, k+3, \ldots, 2k+1, 2k+2\} \cap S$.

Since the binary representation of $2(k+1)$ is exactly the binary representation of $k+1$ attaching a 0 to the end, then $k+1$ and $2k+2$ simultaneously do or do not belong to S. Thus,

$$f(k+1) = \begin{cases} f(k), & \text{if } 2k+1 \notin S, \\ f(k)+1, & \text{if } 2k+1 \in S. \end{cases}$$

We shall show that $\{f(k)\}$ is unbounded, and then together with $f(1) = 0$ claim (a) can be proved.

For positive integer n, the value $f(2n+2)$ is the number of elements of $\{2^n+3, 2^n+4, \ldots, 2^{n+1}+4\} \cap S$. Since $\{2^n, 2^n+1, \ldots, 2^{n+1}-1\}$ is the set of all those numbers whose binary representation has a leading 1 followed by n digits, with exactly two 1s among them, there are C_n^2 ways to choose their positions, and hence the number of elements of $\{2^n, 2^n+1, \ldots, 2^{n+1}-1\} \cap S$ is C_n^2.

Observe that the numbers of 1s in the binary representation of $2^n, 2^n+1$, $2^n+2, 2^{n+1}, 2^{n+1}+1, 2^{n+1}+2, 2^{n+1}+4$ are all less than 3, and the binary representation of $2^{n+1}+3$ has precisely three 1s, so the number of elements of $\{2^n+3, 2^n+4, \ldots, 2^{n+1}+4\} \cap S$ is $C_n^2 + 1$. Then

$$f(2^n+2) = C_n^2 + 1. \tag{$*$}$$

Therefore, $\{f(k)\}$ is unbounded. Claim (a) is proved.

(b) Suppose a positive integer m satisfies that there exists exactly one k with $f(k) = m$. Then from the above discussion,

$$f(k-1) = m-1, \quad f(k+1) = m+1.$$

Also from the conclusions above, $2k-1 \in S$ and $2k+1 \in S$, which implies that both binary representations of $2k-1$ and $2k+1$ have precisely three 1s.

Suppose that the binary representation of k is $(a_n a_{n-1} \ldots a_0)_2$, where $a_n = 1$. Then the binary representation of $2k+1$ is $(a_n a_{n-1} \ldots a_0 1)_2$. Thus there is precisely one 1 in $a_0, a_1, \ldots, a_{n-1}$. Denote it by a_t, and we can see

$$2k - 1 = (2^n + 2^t) \times 2 - 1 = 2^{n+1} + 2^{t+1} - 1,$$

whose binary representation has $1 + t + 1$ ones, so that $t = 1$. Hence $k = 2^n + 2$ ($n \geq 2$). Now by $(*)$ we know m must be $\frac{1}{2}(n^2 - n + 2)$ for some $n \geq 2$. These are all the values we sought for (b).

Remark. For functions defined on positive integers, we can start from studying their recurrence relationship. The calculation in the second half of (a) was prepared for (b). In fact, we can also prove $\{f(k)\}$ is unbounded by other methods.

【Score Situation】 This particular problem saw the following distribution of scores among contestants: 148 contestants scored 7 points, 36 contestants scored 6 points, 19 contestants scored 5 points, 11 contestants scored 4 points, 19 contestants scored 3 points, 18 contestants scored 2 points, 30 contestants scored 1 point, and 104 contestants scored 0 point. The average score of this problem is 3.932, indicating that it was relatively straightforward.

Among the top five teams in the team scores, the scores of this problem are as follows: The United States team scored 42 points (with a total team score of 252 points), the China team scored 42 points (with a total team score of 229 points), the Russia team scored 40 points (with a total team score of 224 points), the Bulgaria team scored 42 points (with a total team score of 223 points), and the Hungary team scored 42 points (with a total team score of 221 points).

The gold medal cutoff for this IMO was set at 40 points (with 30 contestants earning gold medals), the silver medal cutoff was 30 points (with 64 contestants earning silver medals), and the bronze medal cutoff was 19 points (with 98 contestants earning bronze medals).

In this IMO, a total of 22 contestants achieved a perfect score of 42 points.

Problem 1.34 (IMO 55-5, proposed by Luxembourg). For each positive integer n, the Bank of Cape Town issues coins of denomination $\frac{1}{n}$. Given a finite collection of such coins (of not necessarily different denominations) with a total value at most $99 + \frac{1}{2}$, prove that it is possible to split this collection into 100 or fewer groups, such that each group has a total value at most 1.

Proof. We shall prove a general conclusion: for any positive integer N, given a finite collection of such coins with a total value at most $N - \frac{1}{2}$, prove that it is possible to split this collection into N or fewer groups, such that each group has a total value of at most 1.

If some coins have total value of $\frac{1}{k}$ (k is a positive integer), we replace these coins by one coin of value $\frac{1}{k}$, which does not affect the problem. In this way, for each even integer k, at most one coin has value of $\frac{1}{k}$ (otherwise, two such coins may be replaced by one coin of value $\frac{2}{k}$); for each odd integer k, at most $(k-1)$ coins of value $\frac{1}{k}$ (otherwise, k such coins can be replaced by one coin of value 1). So, we may suppose that there is no more replacement that can be made for the coins.

First we take each coin of value 1 as a group. Suppose there are $d < N$ such groups. If there are no other coins, then the problem is solved. Otherwise, take coins of value $1/2$ as group $(d+1)$ if there is any. Let $m = N-d \geq 1$. Then for each integer k in $2, \ldots, m$, take coins of value $\frac{1}{2k-1}$ and value $\frac{1}{2k}$ in group $(d+k)$ if there are any, in which the total value does not exceed $(2k-2) \cdot \frac{1}{2k-1} + \frac{1}{2k} < 1$. For coins of value less than $\frac{1}{(2m)}$, if there are any, we can put them in some group $(d+j)$ such that the total value is less than 1 (since if each group $(d+j)$ has value greater than $1 - \frac{1}{(2m)}$, then the total value will be greater than $d + m(1 - \frac{1}{(2m)}) = m + d - \frac{1}{2} = N - \frac{1}{2}$). Repeat this procedure finitely many times, all coins are put in N or fewer groups with each group of value at most one.

【Score Situation】 This particular problem saw the following distribution of scores among contestants: 84 contestants scored 7 points, 11 contestants scored 6 points, 3 contestants scored 5 points, 8 contestants scored 4 points, 10 contestants scored 3 points, 83 contestants scored 2 points, 60 contestants scored 1 point, and 301 contestants scored 0 point. The average score of this problem is 1.709, indicating that it was relatively challenging.

Among the top five teams in the team scores, the scores of this problem are as follows: The China team scored 35 points (with a total team score of 201 points), the United States team scored 31 points (with a total team score of 193 points), the Chinese Taiwan team scored 29 points (with a total team score of 192 points), the Russia team scored 24 points (with a total team score of 191 points), and the Japan team scored 37 points (with a total team score of 177 points).

The gold medal cutoff for this IMO was set at 29 points (with 49 contestants earning gold medals), the silver medal cutoff was 22 points (with 113 contestants earning silver medals), and the bronze medal cutoff was 16 points (with 133 contestants earning bronze medals).

In this IMO, only three contestants achieved a perfect score of 42 points, namely Jiyang Gao from China, Po-Sheng Wu from Chinese Taiwan and Alexander Gunning from Australia.

Problem 1.35 (IMO 57-3, proposed by Russia). Let $P = A_1 A_2 \cdots A_k$ be a convex polygon in a plane. The vertices $A_1, A_2, \ldots, A_k$ have integral coordinates and lie on a circle. Let S be the area of P. An odd positive integer n is given such that the squares of the side lengths of P are integers divisible by n. Prove that $2S$ is an integer divisible by n.

Proof. By Pick's theorem, $2S$ is an integer. We only need to show, for $n = p^t$ a power of an odd prime, $n \mid 2S$.

We prove by induction on k. When $k = 3$, the polygon P is a triangle, and denote the lengths of its sides as a, b, and c. By the assumption, a^2, b^2, and c^2 are multiples of n. By Heron's formula,

$$16S^2 = 2a^2b^2 + 2b^2c^2 + 2c^2a^2 - a^4 - b^4 - c^4 \equiv 0 (\mod n^2).$$

So, $n \mid 2S$.

Now suppose $k \geq 4$, and the proposition holds for all integers less than k. We claim that P has a diagonal whose length squared is divisible by n. If this is true, we can divide P into two polygons P_1 and P_2 by this diagonal and denote their respective areas S_1 and S_2; by the inductive hypothesis, both $2S_1$ and $2S_2$ are multiples of n, so is $2S = 2S_1 + 2S_2$.

We prove the claim by contradiction. Assume none of the diagonals of P has length squared divisible by $n = p^t$. Let $v_p(N)$ be the exponent of the highest power of p that divides N. Let

$$v_p(A_1 A_m^2) = \alpha = \min v_p(A_i A_j^2) < t,$$

where $2 < m < k$. Apply Ptolemy's theorem to the cyclic quadrilateral $A_1 A_{m-1} A_m A_{m+1}$, and denote $A_1 A_{m-1} = a$, $A_{m-1} A_m = b$, $A_m A_{m+1} = c$, $A_{m+1} A_1 = d$, $A_{m-1} A_{m+1} = e$, $A_1 A_m = f$. Then $ac + bd = ef$. Squaring both sides, we get

$$a^2 c^2 + b^2 d^2 + 2abcd = e^2 f^2.$$

Since a^2, b^2, c^2, d^2, e^2, and f^2 are all integers, so is $2abcd$. Now we analyze the powers of p on both sides.

$$v_p(a^2 c^2) = v_p(c^2) + v_p(a^2) \geq t + \alpha,$$

$$v_p(b^2 d^2) = v_p(b^2) + v_p(d^2) \geq t + \alpha,$$

$$v_p(2abcd) = \frac{1}{2}(v_p(a^2 c^2) + v_p(b^2 d^2)) \geq t + \alpha.$$

So,

$$v_p(a^2c^2 + b^2d^2 + 2abcd) \geq t + \alpha.$$

On the other hand,

$$v_p(e^2 f^2) = v_p(e^2) + v_p(f^2) < t + \alpha,$$

a contradiction. Hence the claim is proved.

Remark. We appreciate the design of this problem, which combines geometry, combination, and number theory very well. To consider n as power of a prime number is not difficult, and the attempt to conduct on k is also natural. However, it is innovative to use Ptolemy's theorem to analyze prime factors. It may be helpful if we focus on the uncommon condition that n points with integral coordinates lie on a circle and try to find some geometric characteristics.

【Score Situation】 This particular problem saw the following distribution of scores among contestants: 10 contestants scored 7 points, 3 contestants scored 6 points, 2 contestants scored 5 points, no contestant scored 4 points, no contestant scored 3 points, 14 contestants scored 2 points, 25 contestants scored 1 point, and 548 contestants scored 0 point. The average score of this problem is 0.251, indicating that it was extremely difficult.

Among the top five teams in the team scores, the scores of this problem are as follows: The United States team scored 27 points (with a total team score of 214 points), the South Korea team scored 23 points (with a total team score of 207 points), the China team scored 20 points (with a total team score of 204 points), the Singapore team scored 11 points (with a total team score of 196 points), and the Chinese Taiwan team scored 7 points (with a total team score of 175 points).

The gold medal cutoff for this IMO was set at 29 points (with 44 contestants earning gold medals), the silver medal cutoff was 22 points (with 101 contestants earning silver medals), and the bronze medal cutoff was 16 points (with 135 contestants earning bronze medals).

In this IMO, a total of six contestants achieved a perfect score of 42 points.

Problem 1.36 (IMO 62-1, proposed by Australia). Let $n \geq 100$ be an integer. Ivan writes the numbers $n, n+1, \ldots, 2n$ each on different cards. He then shuffles these $n + 1$ cards, and divides them into two piles. Prove that at least one of the piles contains two cards such that the sum of their numbers is a perfect square.

Proof. It suffices to find three integers $a, b, c \in [n, 2n]$, such that for some integer k,

$$a + b = (2k - 1)^2, \ a + c = (2k)^2, \ b + c = (2k + 1)^2.$$

In this way, two of a, b, c must be put into one pile, and their sum is a perfect square. Solving the equations yields $a = 2k^2 - 4k$, $b = 2k^2 + 1$, $c = 2k^2 + 4k$. We must require that $n \leq 2k^2 - 4k$ and $2k^2 + 4k \leq 2n$.

Therefore, for a fixed k, the conclusion is true for all integers n in the interval $I_k = [k^2 + 2k, 2k^2 - 4k + 1)$.

When $k \geq 9$, intervals I_k and I_{k+1} are overlapped, since $(k+1)^2 + 2(k+1) \leq 2k^2 - 4k + 1$. Observe that $I_9 \cup I_{10} \cup \cdots = [99, +\infty)$, so the conclusion is true for $n \geq 99$.

Remark. The statement is untrue when $n = 98$. For example, Ivan can put all even numbers between 98 and 126, all odd numbers between 129 and 161, and all even numbers between 162 and 196, into the first pile; he puts the other numbers into the second pile. No two numbers in a pile add up to a perfect square.

【Score Situation】 This particular problem saw the following distribution of scores among contestants: 286 contestants scored 7 points, 36 contestants scored 6 points, 38 contestant scored 5 points, 41 contestants scored 4 points, 10 contestants scored 3 points, 41 contestants scored 2 points, 36 contestants scored 1 point, and 131 contestants scored 0 point. The average score of this problem is 4.394, indicating that it was simple.

Among the top five teams in the team scores, the scores of this problem are as follows: The China team scored 42 points (with a total team score of 208 points), the Russia team scored 40 points (with a total team score of 183 points), the South Korea team scored 40 points (with a total team score of 172 points), the United States team scored 40 points (with a total team score of 165 points), and the Canada team scored 41 points (with a total team score of 151 points).

The gold medal cutoff for this IMO was set at 24 points (with 52 contestants earning gold medals), the silver medal cutoff was 19 points (with 103 contestants earning silver medals), and the bronze medal cutoff was 12 points (with 148 contestants earning bronze medals).

In this IMO, only one contestant achieved a perfect score of 42 points, namely Yichuan Wang from China.

1.3   Summary

It can be observed that in earlier IMOs, there were more number theory problems related to direct divisibility discussions; but now, the problems are more likely to be relatively complex, such as combining with function equations, combinatorics, geometry, and so on.

In the first 64 IMOs, there were a total of 36 divisibility problems. These problems can be broadly categorized into four types, as depicted in Figure 1.1. The score details for these problems are presented in Table 1.2. Due to the smaller number of participating teams and missing contestant score information in early IMOs, there are several blanks in Table 1.2.

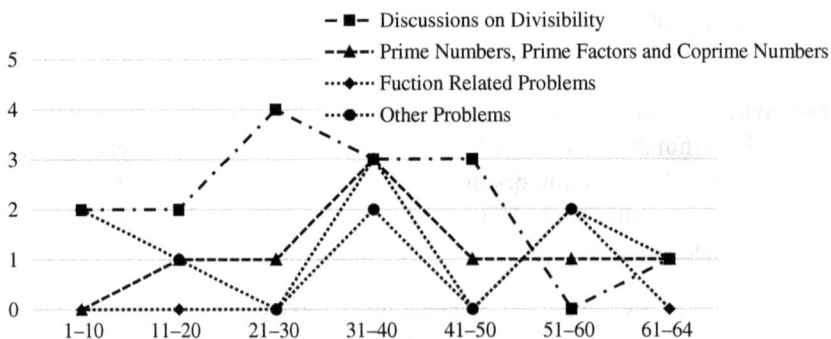

Figure 1.1   Numbers of Divisibility Problems in the First 64 IMOs

Problems 1.1–1.15 focus on "discussions on divisibility"; Problems 1.16–1.23 deal with "prime numbers, prime factors, and coprime numbers"; Problems 1.24–1.28 are about "function related problems"; Problems 1.29–1.36 concern "other problems".

These 36 problems were proposed by 21 countries. Romania contributed six problems; the United Kingdom contributed five problems; Australia contributed three problems; the Netherlands, Poland, Germany, and Bulgaria each contributed two problems.

From Table 1.2, it can be observed that in the first 64 IMOs, there were six divisibility problems with an average score of 0–1 points; four problems with an average score of 1–2 points; eight problems with an average score of 2–3 points; ten problems with an average score of 3–4 points; eight problems with an average score above 4 points. Overall, the divisibility problems had a certain level of difficulty.

Table 1.2 Score Details of Divisibility Problems in the First 64 IMOs

Problem	1.1	1.2	1.3	1.4	1.5	1.6	1.7	1.8
Average score	2.778	3.324	2.333	3.459	1.651	6.275	3.214	1.469
Top five mean			4.775		3.050	6.625	6.033	4.300

Problem	1.9	1.10	1.11	1.12	1.13	1.14	1.15	1.16
Average score	3.649	3.343	3.463	0.591	3.896	4.804	5.845	2.554
Top five mean	6.633	6.567	6.200	2.633	5.900	7.000	7.000	3.725

Problem	1.17	1.18	1.19	1.20	1.21	1.22	1.23	1.24
Average score	3.575	4.221	2.091	1.761	0.778	2.695	2.797	2.399
Top five mean		6.967	5.200	3.767	4.167	6.967	6.400	5.767

Problem	1.25	1.26	1.27	1.28	1.29	1.30	1.31	1.32
Average score	0.815	0.678	0.465	3.259	6.066	5.319	5.943	2.254
Top five mean	3.433	4.167	2.583	6.800				5.233

Problem	1.33	1.34	1.35	1.36
Average score	3.932	1.709	0.251	4.394
Top five mean	6.933	5.200	2.933	6.767

Note. Top five Mean = Total score of the top five teams/Total number of contestants from the top five teams.

Among these 36 problems, the problems with the lowest average score are Problem 1.12 (IMO 43-3, contributed by Romania, discussions on divisibility), Problem 1.21 (IMO 42-6, contributed by Bulgaria, prime numbers, prime factors, and coprime numbers), Problem 1.25 (IMO 38-6, contributed by Lithuania, function related problems), Problem 1.26 (IMO 39-6, contributed by Bulgaria, function related problems), Problem 1.27 (IMO 51-3, contributed by the United States, function related problems), and Problem 1.35 (IMO 57-3, contributed by Russia, other problems). This indicates that comprehensive number theory problems, especially those related to function equations were very difficult.

Chapter 2

Modular Arithmetic

Modular arithmetic is a system concerning the remainders of integers. Broadly speaking, if two integers have the same remainder when divided by the same integer, we call these two integers congruent (specifically, modulo that divisor). The concept of congruence, along with its related notation, was first introduced and extensively used by the renowned German mathematician Gauss.

In high school mathematics competitions, modular arithmetic is a crucial tool for number theory problems. We use the term "tool" here because we consider congruence theory to primarily serve in solving number theory problems. In fact, the remainder itself is a rather elementary mathematical concept. Most students have learnt the complete system of division with quotient and remainder during primary school. The true value of modular arithmetic lies in the statement that two remainders are "the same." Starting from "the same remainder," we can derive a series of related concepts such as "complete residue systems," leading to further theorems like Fermat's little theorem and the Chinese remainder theorem. It can be said that effectively utilizing congruence as a tool can significantly simplify many number theory problems, providing a refreshing perspective.

In the first 64 IMOs, there had been a total of 25 modular arithmetic problems, accounting for approximately 33.3% of all number theory problems. These problems can be primarily categorized into four types: (1) existence problems, totaling 10 problems; (2) finding numbers that satisfy given conditions, totaling eight problems; (3) exploring relationships between terms, totaling four problems; (4) maximum or minimum value problems, totaling three problems. The statistical distribution of these four types of problems in the previous IMOs is presented in Table 2.1.

Table 2.1 Numbers of Modular Arithmetic Problems in the First 64 IMOs

Content	Session							Total
	1–10	11–20	21–30	31–40	41–50	51–60	61–64	
Existence problems	1	2	2	0	3	1	1	10
Finding numbers	0	1	0	3	3	1	0	8
Exploring relationships	0	1	1	0	2	0	0	4
Maximum or minimum values	0	1	0	1	0	1	0	3
Number theory problems	7	10	12	16	14	11	5	75
The percentage of modular arithmetic problems among number theory problems	14.3%	50.0%	25.0%	25.0%	57.1%	27.3%	20.0%	33.3%

It can be observed that modular arithmetic problems are also prevalent in IMO number theory problems, especially in the 11th–20th and the 41st–50th IMOs, where the proportion exceeded 50%. These problems mainly fell into the categories of existence problems and finding numbers that satisfy given conditions.

In fact, there is a close connection between these four types of problems. On the one hand, existence problems mainly involve construction, proving non-existence, etc., while the other three types of problems are based on the assumption that numbers satisfying certain properties already exist. These problems share some similarities in form and methods. On the other hand, they all establish approaches based on congruence "tools." Therefore, rather than focusing solely on specific problems and solutions, it is crucial to appreciate the underlying thought processes reflected in these problems.

This chapter will be divided into three parts. The first part introduces some properties and theorems related to congruence, including Fermat's little theorem, Euler's theorem, Wilson's theorem, the Chinese remainder theorem, etc. These theorems are quite important in number theory problems of high school mathematics competitions.

The second part revolves around four types of problems: "existence problems," "finding numbers that satisfy given conditions," "exploring relationships between terms," and "maximum or minimum value problems." These problems are presented in chronological order, and some problems include various solutions and generalizations.

It is important to note that for each problem, the solutions are followed by information on the scores, including the number of contestants in each score range, the average score, and the scores of the top five teams. However, early IMOs often lacked information on contestant scores, so the number of contestants in each score range only represents the counted number of contestants, and some problems lack scores of the top five teams.

The third part provides a brief summary of this chapter.

2.1 Common Properties, Theorems, and Methods

2.1.1 *Definition and properties of congruence*

(1) *Definition of congruence*

For any integers a and b, if they have the same remainder when divided by a positive integer n, then a and b are said to be congruent modulo n, written as $a \equiv b(\mathrm{mod}\ n)$. Otherwise, we say a and b are not congruent modulo n, written as $a \not\equiv b(\mathrm{mod}\ n)$.

(2) *Properties of congruence*

(i) $a \equiv b(\mathrm{mod}\ n)$ is equivalent to $n|a - b$.

(ii) If $a \equiv b(\mathrm{mod}\ n)$ and $c \equiv d(\mathrm{mod}\ n)$, then $a + c \equiv b + d(\mathrm{mod}\ n)$, $a - c \equiv b - d(\mathrm{mod}\ n)$ and $ac \equiv bd(\mathrm{mod}\ n)$.

(iii) If $a \equiv b(\mathrm{mod}\ n)$ and $a \equiv b(\mathrm{mod}\ m)$, then $a \equiv b(\mathrm{mod}[m, n])$.

(iv) If $ab \equiv ac(\mathrm{mod}\ n)$, then $b \equiv c \left(\mathrm{mod}\frac{n}{(n,a)}\right)$.

(3) *Modular multiplicative inverses*

If $(a, n) = 1$, then there exists integer b, such that $ab \equiv 1(\mathrm{mod}\ n)$. Such b is called a modular multiplicative inverse of a modulo n, denoted as $a^{-1}(\mathrm{mod}\ n)$. In number theory problems, if it is not likely to cause misunderstandings, it can be simply written as a^{-1}.

(4) *Complete residue systems*

Let n be a positive integer. Then the remainder of an integer divided by n must be in the set $\{0, 1, \ldots, n - 1\}$. Thus we can divide the set of integers into n infinite subsets, so that the numbers in the same subset are congruent to each other modulo n. These subsets are called the congruence classes or residue classes modulo n. If we choose one number from each residue class, then we have a set of n integers. This set is called a complete residue system modulo n.

Again let n be a positive integer, a and b are integers, and $(a, n) = 1$. If $x_1, x_2, \ldots, x_n$ form a complete residue system modulo n, then $ax_1 + b, ax_2 + b, \ldots, ax_n + b$ also form a complete residue system modulo n.

(5) *Reduced residue systems*

It is easy to prove that if there exists an integer relatively prime to n in a residue class modulo n, then all the integers in this residue class are relatively prime to n. We say this residue class is relatively prime to n. Choose one number from each residue class relatively prime to n. Then these numbers form a set called reduced residue system modulo n.

2.1.2   *Common theorems and methods for congruence*

(1) *Euler's function*

For positive integer n, the Euler function $\varphi(n)$ denotes the number of positive integers coprime to and less than n. Evidently, the number of residue classes coprime to n and the number of elements in a reduced residue system modulo n are also $\varphi(n)$.

(2) *Fermat's little theorem*

If a is an integer and p is a prime number, then

$$a^p \equiv a \pmod{p}.$$

Especially, if $p \nmid a$, then

$$a^{p-1} \equiv 1 \pmod{p}.$$

Example 2.1. Let $m = \frac{4^p - 1}{3}$ and p is a prime larger than 3. Prove that the remainder of 2^{m-1} divided by m is 1.

Proof.   By Fermat's little theorem, $4^{p-1} \equiv 1 \pmod{p}$. Therefore,

$$3m = 4^p - 1 \equiv 3 \pmod{p}.$$

Since p is a prime larger than 3, we see that $m - 1$ is divisible by p. Then $m - 1 = kp$, where k is a positive integer.

Since $3m = 4^p - 1 = (4 - 1)(4^{p-1} + 4^{p-2} + \cdots + 4 + 1)$, then $m = 4^{p-1} + 4^{p-2} + \cdots + 4 + 1$ is odd, so $m - 1$ is divisible by 2. Suppose

$m - 1 = 2sp$, where k and s are positive integers. Now we have

$$2^{m-1} = 2^{2sp} = 4^{ps} = (3m + 1)^s,$$

so $2^{m-1} \equiv (3m + 1)^s \equiv 1 \pmod{m}$.

As a conclusion, the remainder of 2^{m-1} divided by m is 1.

(3) *Euler's theorem*

If n is a positive integer and a is an integer coprime to n, then

$$a^{\varphi(n)} \equiv 1 \pmod{n}.$$

Example 2.2. We say those positive integers n have property P if for any integer a, as long as $n | a^n - 1$, we have $n^2 | a^n - 1$.

(a) Prove that every prime number n has property P.
(b) Prove that there exist infinitely many composite numbers that have property P.

Proof. (a) Suppose $n = p$ is a prime number and some integer a satisfies $p | (a^p - 1)$. Then $(a, p) = 1$.

By Fermat's little theorem, $p | (a^{p-1} - 1)$. Note that

$$a^p - 1 = a(a^{p-1} - 1) + (a - 1),$$

so $p | (a - 1)$, i.e., $a \equiv 1 \pmod{p}$. Hence

$$a^i \equiv 1 \pmod{p}, i = 0, 1, 2, \ldots, p - 1.$$

We add up the above p congruence equalities to get

$$a^{p-1} + a^{p-2} + \cdots + a + 1 \equiv p \equiv 0 \pmod{p},$$

and then $p^2 | (a - 1)(a^{p-1} + a^{p-2} + \cdots + a + 1)$, i.e., $p^2 | (a^p - 1)$.

(b) Suppose n is a composite number with property P. If $n | (a^n - 1)$, then $(n, a) = 1$.

By Euler's theorem, $a^{\varphi(n)} \equiv 1 \pmod{n}$. Since $a^n \equiv 1 \pmod{n}$, we know

$$a^{(n, \varphi(n))} \equiv 1 \pmod{n}.$$

If $(n, \varphi(n)) = 1$, then $a \equiv 1 \pmod{n}$. Using the proof of (a) we can also conclude that

$$n^2 | (a^n - 1).$$

Therefore, the problem turns to be looking for infinitely many composite numbers n such that

$$(n, \varphi(n)) = 1.$$

For any prime number $p \geq 5$, pick a prime factor q of $p-2$ and let $n = pq$. Now $\varphi(n) = (p-1)(q-1)$. Since $q|(p-2)$, then $q \nmid (p-1)$. Note that $q \leq p-2 < p$, so $p \nmid (q-1)$. Therefore, $(n, \varphi(n)) = 1$.

For such composite number n, if $n|(a^n - 1)$, then $n|(a-1)$, so $a^k \equiv 1(\bmod \, n)$, $k = 0, 1, 2, \ldots, n-1$, and thus $n^2|(a^n - 1)$.

We can find a corresponding composite number $n(p)$ for any prime number $p \geq 5$, and $p < n(p) < p^2$, so there exist infinitely many composite numbers n that have property P.

(4) Wilson's theorem

If and only if p is a prime,

$$(p-1)! \equiv -1(\bmod \, p).$$

Example 2.3. Prove that there does not exist positive integers k, m, such that $k + 1$ is a prime and

$$k! + 48 = 48(k+1)^m.$$

Proof.   We prove it by contradiction. If there exist such positive integers k, m, then since $48|k! + 48$, we have $48|k!$, so $k \geq 6$.

Since $k + 1$ is a prime, by Wilson's theorem,

$$k! \equiv -1 \, (\bmod \, (k+1)),$$

from which $(k+1)|47$, so $k+1 = 47$ (a prime), and thus

$$46! + 48 = 48 \times 47^m.$$

Evidently, $m > 1$, so the above equality is equivalent to

$$\frac{46!}{48} + 1 = 47^m. \tag{1}$$

For this equality, modulo 4 we have $1 \equiv (-1)^m$. Then m is even. Letting $m = 2t$ gives

$$\frac{46!}{48} = (47^t - 1)(47^t + 1).$$

Note that $23^2 \, \big| \, \frac{46!}{48}$. While $47^t + 1 \equiv 2 \, (\bmod \, 23)$, we have $23^2|47^t - 1$.

By the Binomial theorem,

$$47^t = (2 \times 23 + 1)^t \equiv 46t + 1 \pmod{23^2}.$$

Hence, $23|t$, which implies $m = 2t \geq 46$, leading to a contradiction to (1). Therefore, the original proposition is proved.

(5) *Chinese remainder theorem*

Let $m_1, m_2, \ldots, m_k$ be $k(\geq 2)$ coprime positive integers (any pair of them are relatively prime to each other). Now let

$$M = m_1 m_2 \cdots m_k = m_1 M_1 = m_2 M_2 = \cdots = m_k M_k.$$

Then the solutions to the congruence equations

$$\begin{cases} x \equiv b_1 \pmod{m_1}, \\ x \equiv b_2 \pmod{m_2}, \\ \cdots\cdots\cdots\cdots\cdots \\ x \equiv b_k \pmod{m_k} \end{cases}$$

are

$$x \equiv b_1 M_1' M_1 + b_2 M_2' M_2 + \cdots + b_k M_k' M_k \pmod{m},$$

where M_i' is a positive integer that satisfies $M_i' M_i \equiv 1 \pmod{m_i}$, $i = 1, 2, \ldots, k$.

Example 2.4. We divide several students into groups. If five students form a group, then there remains one student; if six students form a group, then there remain five students; if seven students form a group, then there remain four students; if eleven students form a group, then there remain ten students. How many students do we have?

Solution. Suppose the number of students is x. Then by the conditions we have the congruence equations

$$\begin{cases} x \equiv 1 \pmod 5, \\ x \equiv 5 \pmod 6, \\ x \equiv 4 \pmod 7, \\ x \equiv 10 \pmod{11}. \end{cases}$$

This is

$$m_1 = 5, m_2 = 6, m_3 = 7, m_4 = 11;$$
$$b_1 = 1, b_2 = 5, b_3 = 4, b_4 = 10;$$

then
$$M = m_1 m_2 m_3 m_4 = 2310,$$

$$M_1 = \frac{2310}{5} = 462, M_2 = \frac{2310}{6} = 385,$$

$$M_3 = \frac{2310}{7} = 330, M_4 = \frac{2310}{11} = 210.$$

By $1 \equiv M_1' M_1 \equiv 462 M_1' \equiv 2 M_1' (\mathrm{mod}\, 5)$, we have $M_1' = 3$.

By $1 \equiv M_2' M_2 \equiv 385 M_2' \equiv M_2' (\mathrm{mod}\, 6)$, we have $M_2' = 1$.

By $1 \equiv M_3' M_3 \equiv 330 M_1' \equiv M_3' (\mathrm{mod}\, 7)$, we have $M_3' = 1$.

By $1 \equiv M_4' M_4 \equiv 210 M_4' \equiv M_4' (\mathrm{mod}\, 11)$, we have $M_4' = 1$.

Therefore,

$$x \equiv 1 \times 3 \times 462 + 5 \times 1 \times 385 + 4 \times 1 \times 330 + 10 \times 1 \times 210$$

$$= 6731 \equiv 2111 (\mathrm{mod}\, 2310),$$

so $x = 2111 + 2310 t, t = 0, 1, 2, \ldots$.

(6) *Quadratic residues*

Suppose p is a positive integer and n is an integer. If there exists an integer x such that $x^2 \equiv n (\mathrm{mod}\, p)$, then n is called a quadratic residue modulo p. Otherwise, n is called a quadratic non-residue modulo p.

(7) *Legendre symbol*

We use the Legendre symbol $\left(\frac{n}{p}\right)$ to denote the quadratic residue:

If $p | n$, then $\left(\frac{n}{p}\right) = 0$;

If $p \nmid n$ and n is a quadratic residue modulo p, then $\left(\frac{n}{p}\right) = 1$;

If n is a quadratic non-residue modulo p, then $\left(\frac{n}{p}\right) = -1$;

For an odd prime number p, it holds that $\left(\frac{n}{p}\right) = n^{\frac{p-1}{2}}$.

(8) *Law of quadratic reciprocity*

Let p and q be distinct odd prime numbers. Then

$$\left(\frac{q}{p}\right) \cdot \left(\frac{p}{q}\right) = (-1)^{\frac{p-1}{2} \cdot \frac{q-1}{2}}.$$

For the only even prime number 2, we have: 2 is the quadratic residue modulo prime numbers with form $8k \pm 1$, and the quadratic non-residue modulo prime numbers with form $8k \pm 3$.

The law of quadratic reciprocity was first proved in 1796 by Gauss, who referred to it as the "fundamental theorem."

2.2 Problems and Solutions

2.2.1 *Existence problems*

Problem 2.1 (IMO 6-1, proposed by Czechoslovakia). (a) Find all positive integers n for which $2^n - 1$ is divisible by 7.

(b) Prove that there is no positive integer n for which $2^n + 1$ is divisible by 7.

Solution. Since $2^3 = 8 \equiv 1 \pmod{7}$, we can discuss the problem on three cases.

If $n \equiv 0 \pmod 3$, then $n = 3k, k \in \mathbf{N}^*$, so $2^n = (2^3)^k \equiv 1 \pmod 7$. Hence, $2^n - 1$ is divisible by 7 and $2^n + 1$ is not divisible by 7.

If $n \equiv 1 \pmod 3$, then $n = 3k + 1, k \in \mathbf{N}$, so $2^n = 2 \cdot (2^3)^k \equiv 2 \pmod 7$. Hence both $2^n - 1$ and $2^n + 1$ are not divisible by 7.

If $n \equiv 2 \pmod 3$, then $n = 3k + 2, k \in \mathbf{N}$, so $2^n = 4 \cdot (2^3)^k \equiv 4 \pmod 7$. Hence both $2^n - 1$ and $2^n + 1$ are not divisible by 7.

Summarizing, we conclude that $2^n - 1$ is divisible by 7 if and only if n is divisible by 3; for any positive integer n, the positive integer $2^n + 1$ is not divisible by 7.

Remark. A common method for congruence problems is to find the length of the repetend in the exponent. From Fermat's little theorem, when n is coprime to a prime p, we can always find such repetend by $n^k \equiv 1 \pmod p$. In fact, in number theory this is the order of n modulo p.

【Score Situation】 This particular problem saw the following distribution of scores among contestants: 8 contestants scored 7 points, 2 contestants scored 6 points, 1 contestant scored 5 points, 3 contestants scored 4 points, no contestant scored 3 points, 1 contestant scored 2 points, 2 contestants scored 1 point, and no contestant scored 0 point. The average score of this problem is 5.235, indicating that it was simple.

Among the top five teams in the team scores, the scores of this problem are as follows: The Soviet Union team scored 56 points (with a total team score of 269 points), the Hungary team scored 53 points (with a total team score of 253 points), the Romania team scored 50 points (with a total team score of 213 points), the Poland team scored 48 points (with a total team score of 209 points), and the Bulgaria team scored 33 points (with a total team score of 198 points).

The gold medal cutoff for this IMO was set at 38 points (with 7 contestants earning gold medals), the silver medal cutoff was 31 points (with 9 contestants earning silver medals), and the bronze medal cutoff was 27 points (with 19 contestants earning bronze medals).

In this IMO, only one contestant achieved a perfect score of 42 points, namely David Bernstein from the Soviet Union.

Problem 2.2 (IMO 16-3, proposed by Romania). Prove that the number $\sum_{k=0}^{n} 2^{3k} C_{2n+1}^{2k+1}$ is not divisible by 5 for any integer $n \geq 0$.

Proof. Let $x_n = \sum_{k=0}^{n} 2^{3k} C_{2n+1}^{2k+1}$ and $y_n = \sum_{k=0}^{n} 2^{3k} C_{2n+1}^{2k}$. Then

$$(2\sqrt{2} + 1)^{2n+1} = \sum_{k=0}^{2n+1} \left(2^{\frac{3}{2}}\right)^k \cdot C_{2n+1}^{k}$$

$$= \sum_{k=0}^{n} \left(2^{\frac{3}{2}}\right)^{2k+1} C_{2n+1}^{2k+1} + \sum_{k=0}^{n} \left(2^{\frac{3}{2}}\right)^{2k} C_{2n+1}^{2k}$$

$$= 2^{\frac{3}{2}} \sum_{k=0}^{n} 2^{3k} C_{2n+1}^{2k+1} + \sum_{k=0}^{n} 2^{3k} C_{2n+1}^{2k}$$

$$= 2\sqrt{2}x_n + y_n.$$

Similarly, $(2\sqrt{2} - 1)^{2n+1} = 2\sqrt{2}x_n - y_n$. Multiplying these two equalities, we get

$$8x_n^2 - y_n^2 = 7^{2n+1}.$$

Assume for contradiction that $5 | x_n$ for some positive integer n. Then

$$y_n^2 \equiv -7^{2n+1} \equiv -7 \cdot 49^n \equiv -2 \cdot (-1)^n \equiv \pm 2 \,(\text{mod } 5).$$

However, a square number can only be congruent to $0, \pm 1$ modulo 5, which is a contradiction. Therefore the number $\sum_{k=0}^{n} 2^{3k} C_{2n+1}^{2k+1}$ is not divisible by 5 for any non-negative integer n.

Remark. It can easily remind us of the binomial expansion from the given form. However, there are then two difficulties: one is that we cannot eliminate $2^{\frac{3}{2}}$ which is difficult to deal with as an irrational number; the other is the absence of those terms with an even power exponent. To overcome these difficulties, we expanded the binomials of the sum and difference separately to obtain two equalities with only signs different. We could multiply these two equalities and eliminate $2^{\frac{3}{2}}$, and then by discussing congruence modulo 5, the problem is solved.

【Score Situation】 This particular problem saw the following distribution of scores among contestants: 35 contestants scored 8 points, 5 contestants scored 7 points, 1 contestant scored

6 points, 1 contestant scored 5 points, 2 contestants scored 4 points, 9 contestants scored 3 points, 7 contestants scored 2 points, 18 contestants scored 1 point, and 62 contestants scored 0 point. The average score of this problem is 2.807, indicating that it had a certain level of difficulty.

Among the top five teams in the team scores, the scores of this problem are as follows: The Soviet Union team scored 40 points (with a total team score of 256 points), the United States team scored 27 points (with a total team score of 243 points), the Hungary team scored 22 points (with a total team score of 237 points), the German Democratic Republic team scored 42 points (with a total team score of 236 points), and the Yugoslavia team scored 33 points (with a total team score of 216 points).

The gold medal cutoff for this IMO was set at 38 points (with 10 contestants earning gold medals), the silver medal cutoff was 30 points (with 24 contestants earning silver medals), and the bronze medal cutoff was 23 points (with 37 contestants earning bronze medals).

In this IMO, a total of 6 contestants achieved a perfect score of 40 points.

Problem 2.3 (IMO 17-2, proposed by the United Kingdom). Let $a_1, a_2, a_3, \ldots, a_n, \ldots$ be an infinite increasing sequence of positive integers. Prove that for every $p \geq 1$ there are infinitely many a_m which can be written in the form

$$a_m = x a_p + y a_q$$

with x and y positive integers and $q > p$.

Proof. We prove a stronger proposition instead: for any integer $p \geq 1$, there exists an integer $q \geq 1$ (not equal to p), such that there are infinitely many a_m in the sequence which can be written in the form $a_m = x a_p + a_q$.

If $a_p = a_1 = 1$, then let $q = 2$, $m > 2$, and $x = a_m - a_q$.

If $a_p > 1$, then we can divide the original sequence into a_p groups, such that the numbers in the same group are congruent to each other modulo a_p. By the pigeonhole principle, there exists at least one group containing infinitely many numbers. In this group, choose all the numbers exceeding a_p to get a subsequence:

$$a_{m_1} < a_{m_2} < \cdots < a_{m_n} < \cdots .$$

Now, let $q = m_1$. Then for any $m = m_n$, $n \geq 2$, it holds that $a_m > a_q$ and $a_m \equiv a_q \pmod{a_p}$. Thus there exists a positive integer x such that $a_m - a_q = x a_p$. Therefore, the stronger proposition is proved. This solves the original problem.

Remark. The conclusion is relatively weak, since an infinite increasing sequence contains many strong properties. We usually can find something useful in an infinite sequence by the pigeonhole principle and congruence discussion. So after some attempts, we found that one of x and y could be fixed to reach the proof. In fact it does not necessarily have to be fixed as $y = 1$ in our proof. Any fixed positive integer value can be chosen for x or y.

【Score Situation】 This particular problem saw the following distribution of scores among contestants: 65 contestants scored 7 points, 1 contestant scored 6 points, 2 contestants scored 5 points, 4 contestants scored 4 points, 5 contestants scored 3 points, 9 contestants scored 2 points, 13 contestants scored 1 point, and 29 contestants scored 0 point. The average score of this problem is 4.164, indicating that it was simple.

Among the top five teams in the team scores, the scores of this problem are as follows: The Hungary team scored 48 points (with a total team score of 258 points), the German Democratic Republic team scored 56 points (with a total team score of 249 points), the United States team scored 52 points (with a total team score of 247 points), the Soviet Union team scored 48 points (with a total team score of 246 points), and the United Kingdom team scored 51 points (with a total team score of 241 points).

The gold medal cutoff for this IMO was set at 38 points (with 8 contestants earning gold medals), the silver medal cutoff was 32 points (with 25 contestants earning silver medals), and the bronze medal cutoff was 23 points (with 36 contestants earning bronze medals).

In this IMO, a total of six contestants achieved a perfect score of 40 points.

Problem 2.4 (IMO 28-6, proposed by the Soviet Union). Let n be an integer greater than or equal to 2. Prove that if $k^2 + k + n$ is prime for all integers k such that $0 \le k \le \sqrt{\frac{n}{3}}$, then $k^2 + k + n$ is prime for all integers k such that $0 \le k \le n - 2$.

Proof.   Assume for contradiction that m is the smallest integer which makes $m^2 + m + n$ a composite number and $\sqrt{\frac{n}{3}} < m \le n - 2$. Let p be the least prime factor of $m^2 + m + n$. Then

$$p \le \sqrt{m^2 + m + n} \le \sqrt{(n-2)^2 + (n-2) + n} < n.$$

If $m \ge p$, then

$$(m - p)^2 + (m - p) + n \equiv m^2 + m + n \equiv 0 \pmod{p}.$$

Since $(m - p)^2 + (m - p) + n \geq n > p$, then $(m - p)^2 + (m - p) + n$ is composite, which is a contradiction to the definition of m.

If $m \leq p - 1$, then

$$(p - 1 - m)^2 + (p - 1 - m) + n \equiv (m + 1)^2 - (m + 1) + n = m^2 + m + n$$

$$\equiv 0 (\mathrm{mod}\, p).$$

Since $(p-1-m)^2+(p-1-m)+n \geq n > p$, then $(p-1-m)^2+(p-1-m)+n$ is composite. By the definition of m, we have $p - 1 - m \geq m$, so

$$2m + 1 \leq p \leq \sqrt{m^2 + m + n}$$

$$\Rightarrow 4m^2 + 4m + 1 \leq m^2 + m + n$$

$$\Rightarrow 3m^2 + 3m + 1 - n \leq 0$$

$$\Rightarrow m \leq \frac{\sqrt{12n - 3} - 3}{6} < \sqrt{\frac{n}{3}},$$

which is a contradiction to the given condition.

Summarizing, we see that $k^2 + k + n$ is prime for $0 \leq k \leq n - 2$.

Remark. At first glance, it seems difficult to deal with $\sqrt{\frac{n}{3}}$. For such problems, our focus should be on how to find smaller k with the same property from the existing k. This is because as long as we can reduce k continuously, it is likely to fall within the scope by the given condition. For this reason, we started with the least prime factor (as we usually do) and finally obtained a number similar to the form $\sqrt{\frac{n}{3}}$ in the condition.

【Score Situation】 This particular problem saw the following distribution of scores among contestants: 37 contestants scored 7 points, 4 contestants scored 6 points, 5 contestants scored 5 points, 8 contestants scored 4 points, 12 contestants scored 3 points, 12 contestants scored 2 points, 33 contestants scored 1 point, and 126 contestants scored 0 point. The average score of this problem is 1.827, indicating that it was relatively challenging.

Among the top five teams in the team scores, the scores of this problem are as follows: The Romania team scored 42 points (with a total team score of 250 points), the Germany team scored 38 points (with a total team score of 248 points), the Soviet Union team scored 39 points (with a total team score of 235 points), the German Democratic Republic team scored 25 points (with a total team score of 231 points), and the United States team scored 29 points (with a total team score of 220 points).

The gold medal cutoff for this IMO was set at 42 points (with 22 contestants earning gold medals), the silver medal cutoff was 32 points (with 42 contestants earning silver medals), and the bronze medal cutoff was 18 points (with 56 contestants earning bronze medals).

In this IMO, a total of 22 contestants achieved a perfect score of 42 points.

Problem 2.5 (IMO 30-5, proposed by Sweden). Prove that for each positive integer n there exist n consecutive positive integers, none of which is an integral power of a prime number.

Proof 1. Let $N = [(n + 1)!]^2 + 1$. We show that n consecutive positive integers

$$N + 1, N + 2, \ldots, N + n$$

have the demanded property.

This is because for any $N + k(k = 1, 2, \ldots, n)$ of the n integers,

$$N + k = (k + 1)[1^2 \cdot 2^2 \cdot \cdots \cdot k^2 \cdot (k + 1) \cdot (k + 2)^2 \cdot \cdots \cdot (n + 1)^2 + 1].$$

Notice that $k + 1$ is relatively prime to $1^2 \cdot 2^2 \cdot \cdots \cdot k^2 \cdot (k + 1) \cdot (k + 2)^2 \cdot \cdots \cdot (n + 1)^2 + 1$ and both of them are larger than 2, so they are divisible by two different prime divisors separately. Then $N + k$ cannot be an integral power of a prime number.

Proof 2. Take $2n$ different prime numbers. Then by the Chinese remainder theorem, there exist solutions to these congruence equations:

$$x \equiv -1 (\mathrm{mod}\, p_1 q_1),$$

$$x \equiv -2 (\mathrm{mod}\, p_2 q_2),$$

$$\cdots\cdots\cdots\cdots\cdots\cdots\cdots$$

$$x \equiv -n (\mathrm{mod}\, p_n q_n).$$

Suppose a solution N is a positive integer. Then $N + k$ has at least two different prime divisors p_k and q_k for $k = 1, 2, \ldots, n$, so it cannot be an integral power of a prime number. Therefore, $N + 1, N + 2, \ldots, N + n$ have the demanded property.

Remark. This is a construction problem where our goal is to construct n consecutive positive integers with two different prime factors each. One approach is similar to the classical construction for n consecutive composite numbers: $(n+1)!+2, (n+1)!+3, \ldots, (n+1)!+n+1$. To ensure that their prime factors are different, we squared the factorial part. The other method is to set two prime factors for each number directly, and by the Chinese remainder theorem we could ensure that there exist solutions to the congruence equations.

【Score Situation】 This particular problem saw the following distribution of scores among contestants: 142 contestants scored 7 points, 8 contestants scored 6 points, 8 contestants scored 5 points, 2 contestants scored 4 points, 3 contestant scored 3 points, 3 contestants scored 2 points, 3 contestants scored 1 point, and 122 contestants scored 0 point. The average score of this problem is 3.808, indicating that it was relatively straightforward.

Among the top five teams in the team scores, the scores of this problem are as follows: The China team scored 42 points (with a total team score of 237 points), the Romania team scored 42 points (with a total team score of 223 points), the Soviet Union team scored 41 points (with a total team score of 217 points), the German Democratic Republic team scored 42 points (with a total team score of 216 points), and the United States team scored 42 points (with a total team score of 207 points).

The gold medal cutoff for this IMO was set at 38 points (with 20 contestants earning gold medals), the silver medal cutoff was 30 points (with 55 contestants earning silver medals), and the bronze medal cutoff was 18 points (with 72 contestants earning bronze medals).

In this IMO, a total of 10 contestants achieved a perfect score of 42 points.

Problem 2.6 (IMO 41-5, proposed by Russia). Can we find a positive integer n divisible by just 2000 different primes, so that n divides $2^n + 1$? [n may be divisible by a prime power].

Solution. There exists such a positive integer n. We will prove a more general conclusion: for every positive integer k, there exists a positive integer n such that n is a multiple of 3, $n|2^n + 1$, and n is divisible by just k different primes. Then taking $k = 2000$ leads to the original proposition.

First we prove a lemma.

Lemma. For any integer $m > 2$, there exists a prime number p such that $p|m^3 + 1$, but $p \nmid m + 1$.

Proof of the Lemma. Assume to the contrary that for some integer $m > 2$, every prime factor p of $m^2 - m + 1$ divides $m + 1$. Observe that

$$m^2 - m + 1 = (m+1)(m-2) + 3,$$

and then $p|3$, so $p = 3$. Hence $m^2 - m + 1$ is a power of 3.

Since $m + 1 \equiv m - 2 \equiv 0 \pmod 3$, we have

$$m^2 - m + 1 = (m+1)(m-2) + 3 \equiv 3 \pmod 9.$$

As a power of 3, the integer $m^2 - m + 1$ must be 3, which is a contradiction to $m > 2$.

Back to the original problem, when $k = 1$, we see that $n = 3$ meets our requirement.

Suppose the proposition holds for k while n is the demanded integer. Now that n is odd, we obtain

$$2^{2n} - 2^n + 1 = 4^n - 2^n + 1 \equiv 1 - (-1) + 1 \equiv 0 \pmod 3,$$

i.e., $3|2^{2n} - 2^n + 1$. From $2^{3n} + 1 = (2^n + 1)(2^{2n} - 2^n + 1)$, we have $3n|2^{3n} + 1$. By the lemma there exists a prime number p such that $p|2^{3n} + 1$, but $p \nmid 2^n + 1$. Thus $(p, n) = 1$. Then pick $3pn$ for the proposition on case $k + 1$. Now $3pn|2^{3pn} + 1$, and the number of different prime factors of $3pn$ is precisely one more than that of n. With the assumption of the induction hypothesis, $3pn$ is divisible by just $k + 1$ different primes.

Therefore, the proposition holds for $k + 1$. By the method of induction, the proposition holds for every positive integer k. In particular, it holds for $k = 2000$, which is what we need.

Remark. We guessed that this conclusion holds not only for 2000 but also for any positive integer k. So we inducted on k. When there are already k and corresponding n satisfying the conditions, it is a natural idea to multiply n by a new prime number for induction. Therefore, we considered prime numbers in the prime factorization of $2^{Xn} + 1$ that cannot divide n.

【Score Situation】This particular problem saw the following distribution of scores among contestants: 68 contestants scored 7 points, 11 contestants scored 6 points, 1 contestants scored 5 points, 14 contestants scored 4 points, 12 contestants scored 3 points, 16 contestants scored 2 points, 82 contestants scored 1 point, and 257 contestants scored 0 point. The average score of this problem is 1.633, indicating that it was relatively challenging.

Among the top five teams in the team scores, the scores of this problem are as follows: The China team scored 42 points (with a total team score of 218 points), the Russia team scored 42 points (with a total team score of 215 points), the United States team scored 22 points (with a total team score of 184 points), the South Korea team scored 30 points (with a total team score of 172 points), the Bulgaria team scored 30 points (with a total team score of 169 points), and the Vietnam team scored 26 points (with a total team score of 169 points).

The gold medal cutoff for this IMO was set at 30 points (with 39 contestants earning gold medals), the silver medal cutoff was 21 points (with 71 contestants earning silver medals), and the bronze medal cutoff was 11 points (with 119 contestants earning bronze medals).

In this IMO, a total of four contestants achieved a perfect score of 42 points.

Problem 2.7 (IMO 44-6, proposed by France). Show that for each prime p, there exists a prime q such that $n^p - p$ is not divisible by q for any positive integer n.

Proof. Since $\frac{p^p - 1}{p - 1} = 1 + p + p^2 + \cdots + p^{p-1} \equiv p + 1 \pmod{p^2}$, we can get a prime divisor q of $\frac{p^p - 1}{p - 1}$ such that $q \not\equiv 1 \pmod{p^2}$. This q is what we want. The proof is given as follows.

Assume that there exists an integer n such that $n^p \equiv p \pmod{q}$. Then $n^{p^2} \equiv p^p \equiv 1 \pmod{q}$ by the definition of q. On the other hand, from Fermat's little theorem, $n^{q-1} \equiv 1 \pmod{q}$, because q is a prime. Since $p^2 | q - 1$, we have $\gcd(p^2, q - 1) | p$, which leads to $n^p \equiv 1 \pmod{q}$. Hence $p \equiv 1 \pmod{q}$. However, this implies $1 + p + \cdots + p^{p-1} \equiv p \pmod{q}$. From the definition of q, we obtain $0 \equiv \frac{p^p - 1}{p - 1} = 1 + p + \cdots + p^{p-1} \equiv p \pmod{q}$, but this leads to a contradiction.

Remark. If we can find a certain power of some number which is congruent to 1 modulo a prime number, then it is a typical method to combine it with Fermat's little theorem. In fact, the reason we chose such a prime factor q as in the solution is for the later use of Fermat's little theorem.

【Score Situation】This particular problem saw the following distribution of scores among contestants: 24 contestants scored 7 points, 1 contestant scored 6 points, 1 contestant scored 5 points, 0 contestants scored 4 points, 1 contestant scored 3 points, 9 contestants scored 2 points, 64 contestants scored 1 point, and 357 contestants scored 0 point. The average score of this problem is 0.578, indicating that it was extremely difficult.

Among the top five teams in the team scores, the scores of this problem are as follows: The Bulgaria team scored 36 points (with a total team score of 227 points), the China team scored 17 points (with a total team score of 211 points), the United States team scored 14 points (with a total team score of 188 points), the Vietnam team scored 16 points (with a total team score of 172 points), and the Russia team scored 12 points (with a total team score of 167 points).

The gold medal cutoff for this IMO was set at 29 points (with 37 contestants earning gold medals), the silver medal cutoff was 19 points (with 69 contestants earning silver medals), and the bronze medal cutoff was 13 points (with 104 contestants earning bronze medals).

In this IMO, only three contestants achieved a perfect score of 42 points, namely Bào Lê Hùng Việt and Trọng Cành Nguyễn from Vietnam, and Yunhao Fu from China.

Problem 2.8 (IMO 49-3, proposed by Lithuania). Prove that there exist infinitely many positive integers n such that $n^2 + 1$ has a prime divisor which is greater than $2n + \sqrt{2n}$.

Proof 1. Take any integer $m (m \geq 20)$. Suppose p is a prime divisor of $(m!)^2 + 1$. Then $p > m \geq 20$. Take an integer n such that $0 < n < \frac{p}{2}$ and $n \equiv \pm m! \pmod{p}$. Therefore $0 < n < p - n < p$ and

$$n^2 \equiv -1 \pmod{p}. \tag{1}$$

Now

$$(p - 2n)^2 = p^2 - 4pn + 4n^2 \equiv -4 \pmod{p},$$

which yields $(p - 2n)^2 \geq p - 4$,

$$p \geq 2n + \sqrt{p - 4} \geq 2n + \sqrt{2n + \sqrt{p - 4} - 4} > 2n + \sqrt{2n}. \tag{2}$$

Combining (1) and (2), we are done.

Proof 2 (By Zhuo Chen). First, if a prime $p \equiv 1 \pmod 4$, then $\left(\frac{-1}{p}\right) = 1$, i.e., there exists an integer $n \in \{1, 2, \ldots, p-1\}$ such that $n^2 \equiv -1 \pmod p$. Obviously, for this n, $(p - n)^2 \equiv n^2 \equiv -1 \pmod p$ and $\min\{n, p - n\} \leq \frac{p-1}{2}$. Therefore, there exists integer $f(p) \in \{1, 2, \ldots, \frac{p-1}{2}\}$ such that $f^2(p) \equiv -1 \pmod p$.

Next, we show that $n = f(p)$ satisfies

$$2n + \sqrt{2n} < p \text{ if } p \text{ is sufficiently large } (p \geq 29). \qquad (*)$$

Suppose for some $p \geq 29$, the above conclusion is false. Then

$$2n + \sqrt{2n} \geq p \Leftrightarrow (1 + 2\sqrt{2n})^2 \geq 4p + 1,$$

$$\Leftrightarrow n \geq \frac{p-1}{2} - \frac{\sqrt{4p+1} - 3}{4}.$$

Let $t = \frac{p-1}{2} - n \in Z$. Then $0 \leq t \leq \frac{\sqrt{4p+1}-3}{4}$. And

$$0 \equiv n^2 + 1 \equiv \left(\frac{p-1}{2} - t\right)^2 + 1$$

$$= \frac{p^2 - 2p + 1}{4} + t^2 - (p-1)t + 1$$

$$\equiv \frac{3p+5}{4} + t^2 + t \pmod{p}.$$

(Because $4|p - 5$, so $\frac{p^2 - 2p + 1}{4} - \frac{3p+1}{4} = p \cdot \frac{p-5}{4} \equiv 0 \pmod{p}$)

Therefore, $p | t^2 + t + \frac{3p+5}{4}$.

Together with $0 \leq t \leq \frac{\sqrt{4p+1}-3}{4}$, we get

$$0 < t^2 + t + \frac{3p+5}{4} \leq \left(\frac{\sqrt{4p+1}-3}{4}\right)^2 + \frac{\sqrt{4p+1}-3}{4} + \frac{3p+5}{4}$$

$$= \frac{8p + 9 - \sqrt{4p+1}}{8} < p,$$

which is a contradiction.

So the conclusion $(*)$ holds for a sufficiently large prime $p, p \equiv 1 \pmod{4}$.

Finally, it still leaves to show that there are infinity many such $f(p)$. In fact, $p | f^2(p) + 1 \Rightarrow f(p) > \sqrt{p-1}$, and $f(p) \to \infty$ when $p \to \infty$. Therefore, $f(p)$ can take infinitely many values.

As for the original problem, take $n = f(p)$, where p is the corresponding prime divisor.

Remark. This is an existence proving problem. Both proofs roughly found the desired n or p by construction. The first proof started from n, and then by the congruence properties of $(p-2n)^2$ we proved the constructed n meets the conditions. The second proof started from p, which is relatively more straightforward, but requires a brief use of the quadratic residue theory. The subsequent scaling of the inequality is also relatively more difficult.

【Score Situation】This particular problem saw the following distribution of scores among contestants: 46 contestants scored 7 points, 4 contestants scored 6 points, 1 contestant scored 5 points, 3 contestants scored 4 points, 8 contestants scored 3 points, 8 contestants scored 2 points, 27 contestants scored 1 point, and 438 contestants scored 0 point. The average score of this problem is 0.804, indicating that it was extremely difficult.

Among the top five teams in the team scores, the scores of this problem are as follows: The China team scored 42 points (with a total team score of 217 points), the Russia team scored 35 points (with a total team score of 199 points), the United States team scored 20 points (with a total team score of 190 points), the South Korea team scored 26 points (with a total team score of 188 points), and the Iran team scored 14 points (with a total team score of 181 points).

The gold medal cutoff for this IMO was set at 31 points (with 47 contestants earning gold medals), the silver medal cutoff was 22 points (with 100 contestants earning silver medals), and the bronze medal cutoff was 15 points (with 120 contestants earning bronze medals).

In this IMO, only three contestants achieved a perfect score of 42 points, namely Xiaosheng Mou and Dongyi Wei from China, and Alex Zhai from the United States.

Problem 2.9 (IMO 58-6, proposed by the United States). An ordered pair (x, y) of integers is a *primitive point* if the greatest common divisor of x and y is 1. Given a finite set S of primitive points, prove that there exist a positive integer n and integers $a_0, a_1, \ldots, a_n$ such that, for each (x, y) in S,

$$a_0 x^n + a_1 x^{n-1} y + a_2 x^{n-2} y^2 + \cdots + a_{n-1} xy^{n-1} + a_n y^n = 1.$$

Proof 1. First of all, we know that we only need to find a homogeneous polynomial $f(x, y)$ such that, for any point in $S = \{(x_1, y_1), \ldots, (x_n, y_n)\}$, it holds that $f(x_i, y_i) = \pm 1 (i = 1, \ldots, n)$ (then $f^2(x, y)$ meets the requirements). If two of these points are co-linear with the origin $(0, 0)$, then they must be symmetrical about the origin, and the absolute value of any homogeneous polynomial at them is the same. Therefore, we may assume that any two points in S are not collinear with the origin.

Consider the homogeneous polynomial $l_i(x, y) = y_i x - x_i y$. From the definition of a primitive point, $l_i(x_j, y_j) = 0$ if and only if $j = i$. Let $g_i(x, y) = \prod_{j \neq i} l_j(x, y)$. Then $g_i(x, y)$ is a degree $n - 1$ polynomial and has the following two properties:

(1) for $j \neq i$, it is true that $g_i(x_j, y_j) = 0$;

(2) $g_i(x_i, y_i) \neq 0$ (denoted by a_i).

For any integer $N \geq n - 1$, there exists a homogeneous polynomial of degree N which also has the above two properties. Just take a polynomial of degree one $I_i(x, y)$ satisfying $I_i(x_i, y_i) = 1$ (because (x_i, y_i) is a primitive point, such I_i always exists), and then consider $I_i(x, y)^{N-(n-1)} g_i(x, y)$.

Let's reduce the problem to the following proposition.

Proposition. *For any positive integer a, there is an integer coefficient homogeneous polynomial $f_a(x, y)$ of degree no less than 1, so that $f_a(x, y) \equiv 1 \pmod{a}$ for any primitive point (x, y).*

To show that the conclusion of the original problem can be obtained from this proposition, we only need to take a as the least common multiple of the aforementioned $a_i (1 \leq i \leq n)$. Take f_a in the proposition and choose one of its powers $(f_a(x, y))^k$ so that its degree is at least $n - 1$, and then subtract an appropriate integer coefficient linear combination of $g_i(x, y)$ from this polynomial.

Below we prove the proposition by decomposing a. First, when a is a power of a prime number $(a = p^k)$:

- if p is an odd prime number, then let $f_a(x, y) = (x^{p-1} + y^{p-1})^{\varphi(a)}$;
- if $p = 2$, then let $f_a(x, y) = (x^2 + xy + y^2)^{\varphi(a)}$.

Now suppose a is decomposed into $a = q_1 q_2 \ldots q_k$, where q_i is a power of prime numbers and is coprime with each other. Let f_{qi} be a polynomial constructed according to the above rules, and take one of its appropriate powers F_{qi}, so that for all i, the degree of F_{qi} is the same. Note that for any coprime x and y,

$$\frac{a}{q_i} F_{q_i}(x, y) \equiv \frac{a}{q_i} \pmod{a}.$$

According to Bézout's theorem, there exists a linear combination of $\frac{a}{q_i}$ with integer coefficients, whose value is exactly 1. Then the same set of coefficients can be used to form a linear combination of F_{qi} with integer coefficients, so that for any primitive point (x, y), the value of the polynomial is 1 (mod a). Since the degrees of all F_{qi} are the same, we get a homogeneous polynomial.

Proof 2. (Based on the original proof to the problem provided by the contributor, and is slightly simplified by Leader of Israeli Team Dan Carmon).

We use the method of mathematical induction to the number of elements in S. If $|S| = 1$, let $S = \{(x_0, y_0)\}$. According to Bézout's theorem, there are integers a and b such that $ax_0 + by_0 = 1$. Then define an integer coefficient polynomial $P(X, Y) = aX + bY$ such that for any $(x, y) \in S$, we have $P(x, y) = 1$.

Assume $|S| = k \geq 2$, and the conclusion holds for $k - 1$. According to Bézout's theorem, for any $(x_0, y_0) \in S$, there exist integers a and b such that $ax_0 + by_0 = 1$. Now define an integer coefficient linear transformation

$$T : \mathbf{R}^2 \to \mathbf{R}^2, \ T(X, Y) = (aX + bY, -y_0 X + x_0 Y).$$

Then T is also a bijection of $\mathbf{Z}^2$ to $\mathbf{Z}^2$ and maps a primitive point to a primitive point. If there is a homogeneous integer coefficient polynomial $P(X, Y)$, so that $P(x, y) = 1$ for any $(x, y) \in T(S)$, then $P(T(X, Y)) = P(aX + bY, -y_0 X + x_0 Y)$ is also a homogeneous integer coefficient polynomial and satisfies $P(T(x, y)) = 1$ for any $(x, y) \in S$. So we just need to prove the conclusion for $W = T(S)$.

Note that $T(x_0, y_0) = (1, 0) \in W$. Let $W' = W \backslash \{(1, 0)\}$. By induction, there is a homogeneous integer coefficient polynomial $F(X, Y)$, so that $F(x, y) = 1$ for any $(x, y) \in W'$. Suppose $W' = \{(x_1, y_1), \ldots, (x_{k-1}, y_{k-1})\}$.

Let $G(X, Y) = \prod_{i=1}^{k-1}(-x_i Y + y_i X)$. Then $G(x_i, y_i) = 0$ for $1 \leq i \leq k-1$, and $G(1, 0) = y_1 y_2 \ldots y_{k-1} \triangleq a$. Let

$$F(X, Y) = a_0 X^n + a_1 X^{n-1} Y + \cdots + a_n Y^n.$$

Since $F(x_i, y_i) = a_0 x_i^n + y_i(a_1 x_i^{n-1} + \cdots + a_n y_i^{n-1}) = 1$, then $(a_0, y_i) = 1, 1 \leq i \leq n-1$. Therefore $(a_0, a) = 1$. Take a positive integer d such that $a_0^d \equiv 1 \pmod{a}$ and $d > \deg G$. Let $M = \frac{a_0^d - 1}{a} \in \mathbf{Z}$, and

$$P(X, Y) = F(X, Y)^d - M X^{d \deg F - \deg G} G(X, Y).$$

Then $P(X, Y)$ is a homogeneous integer coefficient polynomial with degree $d \deg F$. For $1 \leq i \leq k - 1$,

$$P(x_i, y_i) = F(x_i, y_i)^d - M x_i^{d \deg F - \deg G} G(x_i, y_i) = 1 - 0 = 1$$

and

$$P(1, 0) = F(1, 0)^d - MG(1, 0) - a_0^d - \frac{a_0^d - 1}{a} \cdot a = 1.$$

Remark. As the last problem of the 58th IMO in 2017, neither of these two proofs is easy. The second proof was slightly simplified by Israeli Team leader Dan Carmon based on the proof of the problem provider. We can see that both proofs have an algebra part and a number theory part. Moreover, the number theory part of both proofs uses Bézout's theorem and Euler's theorem for the congruence property.

【Score Situation】 This particular problem saw the following distribution of scores among contestants: 14 contestants scored 7 points, no contestant scored 6 points, 2 contestants scored 5 points, 4 contestants scored 4 points, 5 contestants scored 3 points, 9 contestants scored 2 points, 24 contestants scored 1 point, and 557 contestants scored 0 point. The average score of this problem is 0.294, indicating that it was extremely difficult.

Among the top five teams in the team scores, the scores of this problem are as follows: The South Korea team scored 24 points (with a total team score of 170 points), the China team scored 31 points (with a total team score of 159 points), the Vietnam team scored 14 points (with a total team score of 155 points), the United States team scored 12 points (with a total team score of 148 points), and the Iran team scored 9 points (with a total team score of 142 points).

The gold medal cutoff for this IMO was set at 25 points (with 48 contestants earning gold medals), the silver medal cutoff was 19 points (with 90 contestants earning silver medals), and the bronze medal cutoff was 16 points (with 153 contestants earning bronze medals).

In this IMO, no contestant achieved a perfect score of 42 points.

Problem 2.10 (IMO 63-3, proposed by the United States). Let k be a positive integer and let S be a finite set of odd prime numbers. Prove that there is at most one way (up to rotation and reflection) to place the elements of S around a circle such that the product of any two neighbours is of the form $x^2 + x + k$ for some positive integer x.

Proof. We will allow $x = 0$ in $x^2 + x + k$ and prove the modified statement. Say an unordered prime pair $\{p, q\}(p \neq q)$ is good if there exists a non-negative integer x such that $pq = x^2 + x + k$. We need the following propositions:

(a) for each prime $r \geq 3$, there are at most two odd primes smaller than r, each forming a good pair with r;

(b) if in (a) there are indeed p and q such that (p, r) and (q, r) are good pairs, then (p, q) is also a good pair.

Once (a) and (b) are verified, we induct on $|S|$: for $|S| \leq 3$, the statement is obvious; assume the statement is true for $|S| = n$. Consider $|S| = n+1$: let r be the largest prime in S, and by (a) the neighbors of r around the circle are uniquely determined (up to a reflection); by (b) after removing r the circle of length n is still valid under the problem condition. By the induction hypothesis, there is at most one valid circle of length n. Hence, there is at most one valid circle of length $n+1$, and the modified statement is justified.

Verification of (a). Consider

$$x^2 + x + k \equiv 0 (\bmod r). \qquad (1)$$

According to Lagrange's theorem, (1) has at most two integer solutions $0 \leq x < r$.

Verification of (b). Suppose there exist odd primes p and q with $p < q < r$, and non-negative integers x and y such that

$$x^2 + x + k = pr, \quad y^2 + y + k = qr.$$

As $p < q < r$, it must be $0 \leq x < y \leq r - 1$, and x, y are the two roots of (1). By Vieta's formulas, $x + y \equiv -1(\bmod r)$; from the range of x and y, we infer that $x + y = r - 1$. Let $K = 4k - 1, X = 2x + 1, Y = 2y + 1$. Rewrite the above equalities as

$$4pr = X^2 + K, \quad 4qr = Y^2 + K,$$

where $X + Y = 2r$. Multiply the two equalities to find

$$16pqr = (X^2+K)(Y^2+K) = (XY-K)^2+K(X+Y)^2 = (XY-K)^2+4Kr^2.$$

Hence,

$$4pq = \left(\frac{XY - K}{2r}\right)^2 + K.$$

Since $Z = |\frac{XY-K}{2r}|$ is a rational number and its square $Z^2 = 4pq - K$ is an integer, we deduce that Z itself is an integer. It is easy to see that Z is odd, say $Z = 2z + 1$. Then

$$pq = z^2 + z + k,$$

implying that $\{p, q\}$ is a good pair.

Remark. The construction of a valid cycle appears nontrivial, at least for some k values. For $k = 41$, the following 385 odd primes form a valid cycle: 53, 4357, 104173, 65921, 36383, 99527, 193789, 2089123, 1010357, 2465263, 319169, 15559, 3449, 2647, 1951, 152297, 542189,119773, 91151, 66431, 222137, 1336799, 469069, 45613, 1047941, 656291, 355867,

146669, 874879, 2213327, 305119, 3336209,1623467, 520963, 794201, 1124833, 28697, 15683, 42557, 6571, 39607, 1238833, 835421, 2653681, 5494387, 9357539, 511223,1515317, 8868173, 114079681, 59334071, 22324807, 3051889, 5120939, 7722467, 266239, 693809, 3931783, 1322317, 100469,13913, 74419, 23977, 1361, 62983, 935021, 512657, 1394849, 216259, 45827, 31393, 100787, 1193989, 600979, 209543, 357661, 545141, 19681, 10691, 28867, 165089, 2118023, 6271891, 12626693, 21182429, 1100467, 413089, 772867, 1244423, 1827757,55889, 1558873, 5110711, 1024427, 601759, 290869, 91757, 951109, 452033, 136471, 190031, 4423, 9239, 15809, 24133,115811, 275911, 34211, 877, 6653, 88001, 46261, 317741, 121523, 232439, 379009, 17827, 2699, 15937, 497729, 335539,205223, 106781, 1394413, 4140947, 8346383, 43984757, 14010721, 21133961, 729451, 4997297, 1908223, 278051, 529747,40213, 768107, 456821, 1325351, 225961, 1501921, 562763, 75527, 5519, 9337, 14153, 499, 1399, 2753, 14401, 94583, 245107,35171, 397093, 195907, 2505623, 34680911,18542791, 7415917, 144797293, 455529251, 86675291, 252704911, 43385123,109207907, 204884269, 330414209, 14926789, 1300289, 486769, 2723989, 907757, 1458871, 65063, 4561, 124427, 81343,252887, 2980139, 1496779, 3779057, 519193, 47381, 135283, 268267, 446333, 669481, 22541, 54167, 99439, 158357, 6823,32497, 1390709, 998029, 670343, 5180017, 13936673, 2123491, 4391941, 407651, 209953, 77249, 867653, 427117, 141079,9539, 227, 1439, 18679, 9749, 25453, 3697, 42139, 122327, 712303, 244261, 20873, 52051, 589997, 4310569, 1711069, 291563,3731527, 11045429, 129098443, 64620427, 162661963, 22233269, 37295047, 1936969, 5033449, 725537, 1353973, 6964457, 2176871, 97231, 7001, 11351, 55673, 16747, 169003, 1218571, 479957, 2779783, 949609, 4975787, 1577959, 2365007, 3310753,79349, 23189, 107209, 688907, 252583, 30677, 523, 941, 25981, 205103, 85087, 1011233, 509659, 178259, 950479, 6262847,2333693, 305497, 3199319, 9148267, 1527563, 466801, 17033, 9967, 323003, 4724099, 14278309, 2576557, 1075021, 6462593,2266021, 63922471, 209814503, 42117791, 131659867, 270892249, 24845153, 12104557, 3896003, 219491, 135913, 406397,72269, 191689, 2197697, 1091273, 2727311, 368227, 1911661, 601883, 892657, 28559, 4783, 60497, 31259, 80909, 457697,153733, 11587, 1481, 26161, 15193, 7187, 2143, 21517, 10079, 207643, 1604381, 657661, 126227, 372313, 2176331, 748337,64969, 844867, 2507291, 29317943, 14677801, 36952793, 69332267, 111816223, 5052241, 8479717, 441263, 3020431, 1152751, 13179611, 38280013, 6536771, 16319657, 91442699, 30501409, 49082027, 72061511, 2199433, 167597, 317963, 23869, 2927,3833, 17327, 110879, 285517, 40543, 4861, 21683,

50527, 565319, 277829, 687917, 3846023, 25542677, 174261149, 66370753, 9565711, 1280791, 91393, 6011, 7283, 31859, 8677, 10193, 43987, 11831, 13591, 127843, 358229, 58067, 15473, 65839, 17477,74099, 19603, 82847, 21851, 61.

【Score Situation】 This particular problem saw the following distribution of scores among contestants: 28 contestants scored 7 points, 4 contestants scored 6 points, 4 contestants scored 5 points, 3 contestants scored 4 points, 6 contestants scored 3 points, 69 contestants scored 2 points, 68 contestants scored 1 point, and 407 contestants scored 0 point. The average score of this problem is 0.808, indicating that it was extremely difficult.

Among the top five teams in the team scores, the scores of this problem are as follows: The China team scored 42 points (with a total team score of 252 points), the South Korea team scored 26 points (with a total team score of 208 points), the United States team scored 16 points (with a total team score of 207 points), the Vietnam team scored 19 points (with a total team score of 196 points), and the Romania team scored 17 points (with a total team score of 194 points).

The gold medal cutoff for this IMO was set at 34 points (with 44 contestants earning gold medals), the silver medal cutoff was 29 points (with 101 contestants earning silver medals), and the bronze medal cutoff was 23 points (with 140 contestants earning bronze medals).

In this IMO, a total of 10 contestants achieved a perfect score of 42 points.

2.2.2 *Finding numbers that satisfy given conditions*

Problem 2.11 (IMO 17-4, proposed by the Soviet Union). When 4444^{4444} is written in the decimal notation, the sum of its digits is A. Let B be the sum of the digits of A. Find the sum of the digits of B (A and B are written in decimal notations).

Solution. Since

$$4444^{4444} < (10^4)^{5000} < 10^{20000},$$

the number of digits of 4444^{4444} in the decimal notation is less than 20000. Hence $A \leq 9 \times 20000 = 180000$ and $B \leq 1 + 9 \times 5 = 46$. Suppose the sum of the digits of B is C. Then $C \leq 4 + 9 = 13$.

On the other hand,

$$C \equiv B \equiv A \equiv 4444^{4444} \equiv 7^{4444} \equiv (7^3)^{1481} \cdot 7 \equiv 7 (\mathrm{mod}\ 9).$$

Therefore, it must be $C = 7$, i.e., the sum of the digits of B is 7.

Remark. This problem may be considered difficult in 1975, but today it appears quite simple so that we can even see similar problems in elementary school mathematics competitions. Nevertheless, we should see that scaling and congruence are indeed two powerful tools for dealing with number theory problems.

【Score Situation】 This particular problem saw the following distribution of scores among contestants: 55 contestants scored 6 points, 9 contestants scored 5 points, 6 contestants scored 4 points, 11 contestants scored 3 points, 5 contestants scored 2 points, 5 contestants scored 1 point, and 37 contestants scored 0 point. The average score of this problem is 3.492, indicating that it was relatively straightforward.

Among the top five teams in the team scores, the scores of this problem are as follows: The Hungary team scored 36 points (with a total team score of 258 points), the German Democratic Republic team scored 35 points (with a total team score of 249 points), the United States team scored 42 points (with a total team score of 247 points), the Soviet Union team scored 37 points (with a total team score of 246 points), and the United Kingdom team scored 47 points (with a total team score of 241 points).

The gold medal cutoff for this IMO was set at 38 points (with 8 contestants earning gold medals), the silver medal cutoff was 32 points (with 25 contestants earning silver medals), and the bronze medal cutoff was 23 points (with 36 contestants earning bronze medals).

In this IMO, a total of six contestants achieved a perfect score of 40 points.

Problem 2.12 (IMO 31-3, proposed by Romania). Determine all integers $n > 1$ such that $\frac{2^n+1}{n^2}$ is an integer.

Solution. Obviously n is odd. Let p be the least prime factor of n. Then

$$2^n \equiv -1 \,(\text{mod } p). \tag{1}$$

Let a be the least positive integer satisfying $2^a \equiv -1 \,(\text{mod } p)$, and suppose $n = ka + r$ with $0 \le r < a$. Then

$$2^n \equiv 2^{ka+r} \equiv (-1)^k \cdot 2^r \,(\text{mod } p). \tag{2}$$

If k is even, by (1) and (2), $2^r \equiv -1 \,(\text{mod } p)$. Since a is the least positive integer satisfying this equation, $r = 0$. If k is odd, by (1) and (2), $2^r \equiv 1 \,(\text{mod } p)$, so $2^{a-r} \equiv -1 \,(\text{mod } p)$. With $0 < a - r \le a$ and the definition of a, we also have $r = 0$.

Therefore, it always holds that $n = ka$. By Fermat's little theorem, $2^{p-1} \equiv 1 \pmod{p}$. If $a \geq p$, then $0 < a - (p-1) < a$ and $2^{a-(p-1)} \equiv -1 \pmod{p}$, which is a contradiction to the definition of a. Hence $a \leq p-1$.

Since p is the least prime factor of n and $a|n$, we have $a < p$, so $a = 1$, which implies $2^1 \equiv -1 \pmod{p}$, and thus $p = 3$.

Suppose $n = 3^k d$, where $k \geq 1$, $(d,3) = 1$, and d is odd. Assume that $k \geq 2$. Then

$$2^n + 1 = (3-1)^n + 1 = 3n - \sum_{j=2}^{n} (-1)^j C_n^j 3^j. \tag{3}$$

Since the exponent of 3 in $j!$ is

$$\left[\frac{j}{3}\right] + \left[\frac{j}{3^2}\right] + \cdots < \frac{j}{3} + \frac{j}{3^2} + \cdots = \frac{j}{2},$$

the exponent of 3 in $C_n^j = \frac{n(n-1)\cdots(n-j+1)}{j!}$ must be $> k - \frac{j}{2}$. Then for $j \geq 2$, the exponent of 3 in $(-1)^j C_n^j 3^j$ is not less than $k - \frac{j}{2} + j + 1 = k + \frac{j}{2} + 1$. Hence $\sum_{j=2}^{n} (-1)^j C_n^j 3^j$ is divisible by 3^{k+2}. Note that $3n$ is divisible by 3^{k+1} but not divisible by 3^{k+2}, so $2^n + 1$ cannot be divisible by 3^{k+2}. However $n^2 = 3^{2k} d^2$ and $2k \geq k + 2$, so $n^2 \nmid 2^n + 1$, a contradiction. Therefore, $n = 3d$ with $(d,3) = 1$.

Assume that $d > 1$, and let q be the least prime factor of d. Then $q \geq 5$ and

$$2^n \equiv -1 \pmod{q}.$$

Suppose b is the least positive integer such that $2^b \equiv -1 \pmod{q}$. Similar to the previous discussion, we know that $j|n$ and $j \leq q - 1$. Hence $j = 1$ or 3, and correspondingly, $q|3$ or $q|9$, both are contradictory to $q \geq 5$. Thus it must be $d = 1$ and $n = 3$.

We can easily find that $n = 3$ satisfies the given condition, so the solution is $n = 3$.

Remark. This is another problem analyzing least prime factors by Fermat's little theorem.

【Score Situation】 This particular problem saw the following distribution of scores among contestants: 16 contestants scored 7 points, 6 contestants scored 6 points, 6 contestants scored 5 points, 16 contestants scored 4 points, 42 contestant scored 3 points, 79 contestants scored 2 points, 118 contestants scored 1 point, and 25 contestants scored 0 point. The average score of this problem is 2.091, indicating that it had a certain level of difficulty.

Among the top five teams in the team scores, the scores of this problem are as follows: The China team scored 35 points (with a total team score of 230 points), the Soviet Union team scored 26 points (with a total team score of 193 points), the United States team scored 20 points (with a total team score of 174 points), the Romania team scored 26 points (with a total team score of 171 points), and the France team scored 22 points (with a total team score of 168 points).

The gold medal cutoff for this IMO was set at 34 points (with 23 contestants earning gold medals), the silver medal cutoff was 23 points (with 56 contestants earning silver medals), and the bronze medal cutoff was 16 points (with 76 contestants earning bronze medals).

In this IMO, a total of four contestants achieved a perfect score of 42 points.

Problem 2.13 (IMO 36-6, proposed by Poland). Let p be an odd prime number. How many p-element subsets A of $\{1, 2, \ldots, 2p\}$ are there, the sum of whose elements is divisible by p?

Solution. Let $W = \{1, 2, \ldots, 2p\}$, $U = \{1, 2, \ldots, p\}$, and $V = \{p+1, p+2, \ldots, 2p\}$. Then U and V are two subsets satisfying the given conditions.

Evidently, the number of p-elements subsets of W is C_{2p}^p. For any p-element subset A (except U and V), the intersection of A and U, or A and V is not empty. Now we divide these subsets into several groups such that two p-element subsets (except U and V) S and T are in the same group if and only if both of these two conditions hold:

(a) $S \bigcap V = T \bigcap V$;

(b) there exists a way to number the elements $s_1, s_2, \ldots, s_m$ of $S \bigcap U$ and the elements $t_1, t_2, \ldots, t_m$ of $T \bigcap U$, such that for some $k \in \{0, 1, 2, \ldots, p-1\}$, the congruence equations $s_i - t_i \equiv k \pmod{p}$ with $i = 1, 2, \ldots, m$ hold.

Then for distinct k_1 and k_2 in $\{0, 1, 2, \ldots, p-1\}$, suppose T_1 and T_2 are the corresponding sets getting from a same S by (b). Then calculating the sum of their elements, we have

$$\sum_{i=1}^{p} s_i - \sum_{i=1}^{p} t_i \equiv (k_1 - k_2)m \pmod{p},$$

where m is the number of elements of $S \bigcap U$. Since $S \bigcap U \neq \varnothing$, we have $0 < m < p$, and then $(k_1 - k_2)m \not\equiv 0 \pmod{p}$, which implies $T_1 \neq T_2$. So there are precisely p subsets in every group, the sums of whose elements are not congruent to each other modulo p. Hence, there is precisely one subset in every group satisfying the original condition.

As a result, we can divide all the C_{2p}^p p-element subsets of W, excluding U and V, into several groups, such that each group contains p subsets and among these p subsets there is exactly one subset we want. Therefore, the number of such subsets is $\frac{1}{p}(C_{2p}^p - 2) + 2$.

Remark. The background of this problem is actually an application of the complete residue system although we did not actually mention it.

【Score Situation】 This particular problem saw the following distribution of scores among contestants: 34 contestants scored 7 points, 6 contestants scored 6 points, 5 contestants scored 5 points, 5 contestant scored 4 points, 12 contestant scored 3 points, 22 contestants scored 2 points, 37 contestants scored 1 point, and 291 contestants scored 0 point. The average score of this problem is 1.058, indicating that it was relatively challenging.

Among the top five teams in the team scores, the scores of this problem are as follows: The China team scored 34 points (with a total team score of 236 points), the Romania team scored 30 points (with a total team score of 230 points), the Russia team scored 17 points (with a total team score of 227 points), the Vietnam team scored 12 points (with a total team score of 220 points), and the Hungary team scored 27 points (with a total team score of 210 points).

The gold medal cutoff for this IMO was set at 37 points (with 30 contestants earning gold medals), the silver medal cutoff was 29 points (with 71 contestants earning silver medals), and the bronze medal cutoff was 19 points (with 100 contestants earning bronze medals).

In this IMO, a total of 14 contestants achieved a perfect score of 42 points.

Problem 2.14 (IMO 40-4, proposed by Chinese Taiwan). Determine all pairs (n, p) of positive integers such that p is a prime, n is not exceeded $2p$, and $(p-1)^n + 1$ is divisible by n^{p-1}.

Solution. When $n = 1$, obviously p is divisible by 1 for any prime p.

When $p = 2$, we want 2 to be divisible by n, which leads to $n = 1$ or 2.

Next we consider the cases when $n \geq 2$ and $p \geq 3$. Now that $(p-1)^n + 1$ is odd, n is also odd and $n < 2p$. Let q be the least prime divisor of n. Then q is odd, and $(n, q-1) = 1$. By Bézout's theorem, there exist integers

u and v such that

$$un + v(q-1) = 1.$$

As $q-1$ is even, u must be odd.

By the given condition,

$$(p-1)^n \equiv -1 (\mod q);$$

on the other hand, by Fermat's little theorem,

$$(p-1)^{q-1} \equiv 1 (\mod q).$$

Then by these two equalities,

$$p - 1 = (p-1)^{un+v(q-1)}$$
$$= (p-1)^{un} \cdot (p-1)^{v(q-1)}$$
$$\equiv (-1)^u \cdot 1^v$$
$$= -1 (\mod q).$$

This is to say $q|p$, so $q = p$. From the definition of q we have $p|n$, together with $n < 2p$ we know $n = p$. Hence $p^{p-1}|((p-1)^p + 1)$.

Since

$$(p-1)^p + 1 = p^2(p^{p-2} - C_p^1 p^{p-3} + \cdots - C_p^{p-2} + 1),$$

while

$$p^{p-2} - C_p^1 p^{p-3} + \cdots - C_p^{p-2} + 1 \equiv 1 (\mod p),$$

we conclude that $p - 1 \le 2$, which means $p = 3$. By checking $n = p = 3$ we can see this pair of (n, p) meets our requirements.

Summarizing, we see that all wanted pairs (n, p) are $(n, p) = (2, 2), (3, 3), (1, p)$, where p is any prime.

【Score Situation】 This particular problem saw the following distribution of scores among contestants: 75 contestants scored 7 points, 9 contestants scored 6 points, 15 contestants scored 5 points, 24 contestants scored 4 points, 59 contestants scored 3 points, 109 contestants scored 2 points, 119 contestants scored 1 point, and 40 contestants scored 0 point. The average score of this problem is 2.809, indicating that it had a certain level of difficulty.

Among the top five teams in the team scores, the scores of this problem are as follows: The China team scored 42 points (with a total team score of 182 points), the Russia team scored 34 points (with a total team score of 182 points), the Vietnam team scored 32 points (with a total team score of 177 points), the Romania team scored 31 points (with a total team score of 173 points), and the Bulgaria team scored 42 points (with a total team score of 170 points).

The gold medal cutoff for this IMO was set at 28 points (with 38 contestants earning gold medals), the silver medal cutoff was 19 points (with 70 contestants earning silver medals), and the bronze medal cutoff was 12 points (with 118 contestants earning bronze medals).

In this IMO, no contestant achieved a perfect score of 42 points.

Problem 2.15 (IMO 45-6, proposed by Iran). We call a positive integer *alternating* if every two consecutive digits in its decimal representation are of different parity.

Find all positive integers n such that n has a multiple which is alternating.

Solution First we prove two lemmas.

Lemma 1. *If k is a positive integer, then there exist $0 \leq a_1, a_2, \ldots, a_{2k} \leq 9$ such that $a_1, a_3, \ldots, a_{2k-1}$ are odd integers, $a_2, a_4, \ldots, a_{2k}$ are even integers, and*

$$2^{2k+1} | \overline{a_1 a_2 \cdots a_{2k}} \text{ (The decimal representation).}$$

Proof of Lemma 1. By mathematical induction.

If $k = 1$, it follows from $8 \mid 16$ that the proposition is true.

Assume that for $k = n - 1$, the proposition is true.

When $k = n$, let

$$\overline{a_1 a_2 \cdots a_{2n-2}} = 2^{2n-1} t$$

by the inductive hypothesis. The problem reduces to proving that there exist $1 \leq a, b \leq 9$ with a odd and b even such that $2^{2n+1} | \overline{ab} \times 10^{2n-2} + 2^{2n-1}t$, or $8 | \overline{ab} \times 5^{2n-2} + 2t$, or $8 | \overline{ab} + 2t$ in view of $5^{2n-2} \equiv 1 \pmod 8$. It follows from $8 | 12 + 4, 8 | 14 + 2, 8 | 16 + 0$, and $8 | 50 + 6$ that Lemma 1 is true.

Lemma 2. *If k is a positive integer, then there exists an alternative number $\overline{a_1 a_2 \cdots a_{2k}}$ with an even number $2k$ of digits such that a_{2k} is odd and $5^{2k} | \overline{a_1 a_2 \cdots a_{2k}}$, where a_1 can be 0, but $a_2 \neq 0$.*

Proof of Lemma 2. By mathematical induction.

If $k = 1$, it follows from $25 \mid 25$ that the proposition is true.

Assume that for $k = n - 1$, the proposition is true, or there exists an alternative multiple of $\overline{a_1 a_2 \cdots a_{2k}}$ satisfying $5^{2k} \mid \overline{a_1 a_2 \cdots a_{2k}}$.

When $k = n$, let

$$\overline{a_1 a_2 \cdots a_{2n-2}} = 5^{2n-1} t.$$

The problem reduces to proving that there exist $0 \leq a, b \leq 9$ with a even and b odd such that $5^{2n} \mid \overline{ab} \times 10^{2n-2} + 5^{2n-2} t$ or $25 \mid \overline{ab} \times 2^{2n-2} + t$.

Since 2^{2n-2} is coprime to 25, there exist a and b such that $0 < \overline{ab} \leq 25$ and $25 \mid \overline{ab} \times 2^{2n-2} + t$. If b is odd, then between $\overline{ab}$ and $\overline{ab} + 50$, at least one satisfies that the highest-valued digit is even. If b is even, then between $\overline{ab} + 25$ and $\overline{ab} + 75$, at least one satisfies that the highest-valued digit is even. This completes the proof of Lemma 2.

Back to the original problem. Let $n = 2^\alpha 5^\beta t$, where t is coprime to 10 and $\alpha, \beta \in \mathbb{N}$. Assume that $\alpha \geq 2, \beta \geq 1$. Let l be an arbitrary multiple of n. The last decimal digit is 0, and the digit in tens is even. Hence these n do not satisfy the required condition.

(1) When $\alpha = \beta = 0$, consider $21, 2121, 212121, \ldots, \underbrace{2121 \cdots 21}_{\text{number } k \text{ of } 21}, \ldots$.

There must exist two of them congruent modulo n. Without loss of generality, we may assume that $t_1 > t_2$, and

$$\underbrace{2121 \cdots 21}_{\text{number } t_1 \text{ of } 21} \equiv \underbrace{2121 \cdots 21}_{\text{number } t_2 \text{ of } 21} \pmod{n}.$$

Then

$$\underbrace{2121 \cdots 21}_{\text{number } t_1 - t_2 \text{ of } 21} \underbrace{00 \cdots 0}_{\text{number } 2t_2 \text{ of } 0} \equiv 0 \pmod{n}.$$

Hence

$$n \mid \underbrace{2121 \cdots 21}_{\text{number } t_1 - t_2 \text{ of } 21},$$

because n is coprime to 10.

Now these positive integers n satisfy the required condition.

(2) When $\beta = 0$ and $\alpha \geq 1$, it follows from Lemma 1 that there exists an alternative number $\overline{a_1 a_2 \cdots a_{2k}}$ that is a multiple of 2^α. Consider

$$\overline{a_1 a_2 \cdots a_{2k}}, \overline{a_1 a_2 \cdots a_{2k} a_1 a_2 \cdots a_{2k}}, \ldots,$$

$$\underbrace{\overline{a_1 a_2 \cdots a_{2k} a_1 a_2 \cdots a_{2k} \cdots a_1 a_2 \cdots a_{2k}}}_{\text{number } l \text{ of sections}}, \ldots,$$

there must exist two of them congruent modulo t. Without loss of generality, we may assume that $t_1 > t_2$, and

$$\underbrace{\overline{a_1 a_2 \cdots a_{2k} a_1 a_2 \cdots a_{2k} \cdots a_1 a_2 \cdots a_{2k}}}_{\text{number } t_1 \text{ of sections}}$$

$$\equiv \underbrace{\overline{a_1 a_2 \cdots a_{2k} a_1 a_2 \cdots a_{2k} \cdots a_1 a_2 \cdots a_{2k}}}_{\text{number } t_2 \text{ of sections}} (\text{mod } t).$$

Then

$$\underbrace{\overline{a_1 a_2 \cdots a_{2k} a_1 a_2 \cdots a_{2k} \cdots a_1 a_2 \cdots a_{2k}}}_{\text{number } t_1 - t_2 \text{ of sections}} \underbrace{\overline{00 \cdots 0}}_{\text{number } 2kt_2 \text{ of } 0} \equiv 0 (\text{mod } t).$$

Since t is coprime to 10, $t | \underbrace{\overline{a_1 a_2 \cdots a_{2k} a_1 a_2 \cdots a_{2k} \cdots a_1 a_2 \cdots a_{2k}}}_{\text{number } t_1 - t_2 \text{ of sections}}$. Moreover, since t is coprime to 2,

$$n = 2^\alpha t | \underbrace{\overline{a_1 a_2 \cdots a_{2k} a_1 a_2 \cdots a_{2k} \cdots a_1 a_2 \cdots a_{2k}}}_{\text{number } t_1 - t_2 \text{ of sections}},$$

which is alternating.

(3) When $\alpha = 0, \beta \geq 1$, it follows from Lemma 2 that there exists an alternative multiple that $\overline{a_1 a_2 \cdots a_{2k}}$ is a multiple of 5^β with a_{2k} odd.

Using the same argument as in (2), we obtain that there exist $t_1 > t_2$ satisfying $t | \underbrace{\overline{a_1 a_2 \cdots a_{2k} a_1 a_2 \cdots a_{2k} \cdots a_1 a_2 \cdots a_{2k}}}_{\text{number } t_1 - t_2 \text{ of sections}}$. Since t is coprime to 5, $5^\beta t | \underbrace{\overline{a_1 a_2 \cdots a_{2k} a_1 a_2 \cdots a_{2k} \cdots a_1 a_2 \cdots a_{2k}}}_{\text{number } t_1 - t_2 \text{ of sections}}$. In addition, $\underbrace{\overline{a_1 a_2 \cdots a_{2k} a_1 a_2 \cdots a_{2k} \cdots a_1 a_2 \cdots a_{2k}}}_{\text{number } t_1 - t_2 \text{ of sections}}$ is alternating, and the last decimal digit a_{2k} is odd.

(4) When $\alpha = 1, \beta \geq 1$, it follows from (3) that there exists an alternative number $\overline{a_1 a_2 \cdots a_{2k} \cdots a_1 a_2 \cdots a_{2k}}$ satisfying that a_{2k} is odd and

$$5^\beta t | \overline{a_1 a_2 \cdots a_{2k} \cdots a_1 a_2 \cdots a_{2k}}.$$

Hence $n = 2 \cdot 5^\beta t | \overline{a_1 a_2 \cdots a_{2k} \cdots a_1 a_2 \cdots a_{2k} 0}$, which is alternating.

In conclusion, if n is not divisible by 20, then these positive integers n satisfy the required condition.

Remark. When discussing problems related to decimal digits, an important point is to check the properties modulo 2 and 5. In this proof, both lemmas constructed the desired integer by mathematical induction. For the original problem, we repeated a section for sufficiently many times to find two terms congruent modulo t, and then the difference between these two terms has the congruence properties we want. The background of this method is the pigeonhole principle. Similar methods are commonly used for problems related to complete residue systems.

【Score Situation】 This particular problem saw the following distribution of scores among contestants: 48 contestants scored 7 points, 4 contestants scored 6 points, 6 contestants scored 5 points, 7 contestants scored 4 points, 14 contestants scored 3 points, 33 contestants scored 2 points, 85 contestants scored 1 point, and 289 contestants scored 0 point. The average score of this problem is 1.257, indicating that it was relatively challenging.

Among the top five teams in the team scores, the scores of this problem are as follows: The China team scored 42 points (with a total team score of 220 points), the United States team scored 34 points (with a total team score of 212 points), the Russia team scored 27 points (with a total team score of 205 points), the Vietnam team scored 37 points (with a total team score of 196 points), and the Bulgaria team scored 19 points (with a total team score of 194 points).

The gold medal cutoff for this IMO was set at 32 points (with 45 contestants earning gold medals), the silver medal cutoff was 24 points (with 78 contestants earning silver medals), and the bronze medal cutoff was 16 points (with 120 contestants earning bronze medals).

In this IMO, a total of four contestants achieved a perfect score of 42 points.

Problem 2.16 (IMO 46-4, proposed by Poland). Determine all positive integers relatively prime to all the terms of the infinite sequence

$$a_n = 2^n + 3^n + 6^n - 1 \ (n = 1,\, 2,\, 3,\, \dots).$$

Solution. First, we prove the following result: for a fixed prime $p(p \geq 5)$,

$$2^{p-2} + 3^{p-2} + 6^{p-2} - 1 \equiv 0 \pmod{p}. \tag{1}$$

Since p is a prime not less than 5, we have $(2,p) = (3,p) = (6,p) = 1$. By Fermat's little theorem,

$$2^{p-1} \equiv 1 \pmod p, \quad 3^{p-1} \equiv 1 \pmod p, \quad 6^{p-1} \equiv 1 \pmod p.$$

Therefore

$$3 \cdot 2^{p-1} + 2 \cdot 3^{p-1} + 6^{p-1} \equiv 3 + 2 + 1 = 6 \pmod p,$$

i.e.,

$$6 \cdot 2^{p-2} + 6 \cdot 3^{p-2} + 6 \cdot 6^{p-2} \equiv 6 \pmod p.$$

Simplifying gives

$$2^{p-2} + 3^{p-2} + 6^{p-2} - 1 \equiv 0 \pmod p.$$

So (1) holds, and $a_{p-2} = 2^{p-2} + 3^{p-2} + 6^{p-2} - 1 \equiv 0 \pmod p$.

It is trivial that $a_1 = 10$, $a_2 = 48$.

For any integer n greater than 1, it has a prime factor p. If $p \in \{2, 3\}$, then $(n, a_2) > 1$. If $p \geq 5$, then $(n, a_{p-2}) > 1$. Therefore we can claim that every integer greater than 1 does not match the condition.

Since 1 is relatively prime to every other positive integer, 1 is the only number satisfying the condition.

Remark. It is not difficult to realize that whether a positive integer is coprime to every term in the sequence depends on whether its prime factors satisfy this condition. The reason why we came up with the idea of using Fermat's little theorem and taking the exponent as $p-2$ is the familiar equality $\frac{1}{2} + \frac{1}{3} + \frac{1}{6} = 1$. In fact, if we know the concept of modular multiplicative inverse, then this equality also holds modulo p.

【Score Situation】 This particular problem saw the following distribution of scores among contestants: 238 contestants scored 7 points, 7 contestants scored 6 points, 3 contestants scored 5 points, no contestant scored 4 points, no contestant scored 3 points, 30 contestants scored 2 points, 145 contestants scored 1 point, and 90 contestants scored 0 point. The average score of this problem is 3.758, indicating that it was relatively straightforward.

Among the top five teams in the team scores, the scores of this problem are as follows: The China team scored 42 points (with a total team score of 235 points), the United States team scored 42 points (with a total team score of 213 points), the Russia team scored 42 points (with a total team score of 212 points), the Iran team scored 42 points (with a total team score of 201 points), and the South Korea team scored 42 points (with a total team score of 200 points).

The gold medal cutoff for this IMO was set at 35 points (with 42 contestants earning gold medals), the silver medal cutoff was 23 points (with 79 contestants earning silver medals), and the bronze medal cutoff was 12 points (with 128 contestants earning bronze medals).

In this IMO, a total of 16 contestants achieved a perfect score of 42 points.

Problem 2.17 (IMO 48-5, proposed by the United Kingdom). Let a and b be positive integers. Show that if $4ab - 1$ divides $(4a^2 - 1)^2$, then $a = b$.

Proof. Call (a, b) a "bad pair" if it satisfies $4ab - 1 \mid (4a^2 - 1)^2$ while $a \neq b$. We use the infinite descent method to prove that there is no such "bad pair."

Property (1). If (a, b) is a 'bad pair' and $a < b$, then there exists an integer c $(c < a)$ such that (a, c) is also a "bad pair."

In fact, let $r = \frac{(4a^2 - 1)^2}{4ab - 1}$. Then

$$r = -r \cdot (-1) \equiv -r(4ab - 1)$$

$$= -(4a^2 - 1)^2 \equiv -1 (\mathrm{mod}\ 4a).$$

Therefore there exists an integer c such that $r = 4ac - 1$. Since $a < b$, we have $4ac - 1 = \frac{(4a^2 - 1)^2}{4ab - 1} < 4a^2 - 1$. So $c < a$ and $4ac - 1 \mid (4a^2 - 1)^2$. Thus (a, c) is a "bad pair" too.

Property (2). If (a, b) is a "bad pair," then so is (b, a).

In fact, by $1 = 1^2 \equiv (4ab)^2 (\mathrm{mod}\ (4ab - 1))$, we get

$$(4b^2 - 1) \equiv (4b^2 - (4ab)^2) = 16b^4(4a^2 - 1)$$

$$\equiv 0 (\mathrm{mod}\ (4ab - 1)).$$

Thus, $4ab - 1 \mid (4b^2 - 1)$.

In the following we will show such "bad pair" does not exist. We prove it by contradiction.

Suppose there is at least one "bad pair." We choose such a pair for which $2a + b$ is the least. If $a < b$, then by property (1), there is a "bad pair" (a, c) which satisfies $c < b$, and $2a + c < 2a + b$, a contradiction. If $b < a$, then by property (2), (b, a) is also a "bad pair," leads to $2b + a < 2a + b$, a contradiction.

All these show that such a "bad pair" does not exist. Therefore $a = b$.

Remark. The main idea is similar to a previous problem using Vieta's formula for constantly finding smaller solutions. In the latter half, we can also choose such a pair (a, b) for which $a + b$ is the least instead of $2a + b$. If $a < b$, then the contradiction can be obtained from property (1); or $a > b$. Using property (2) and then property (1), we can also obtain the contradiction.

【Score Situation】 This particular problem saw the following distribution of scores among contestants: 94 contestants scored 7 points, 6 contestants scored 6 points, 4 contestants scored 5 points, 3 contestants scored 4 points, 10 contestants scored 3 points, 38 contestants scored 2 points, 155 contestants scored 1 point, and 210 contestants scored 0 point. The average score of this problem is 1.898, indicating that it was relatively challenging.

Among the top five teams in the team scores, the scores of this problem are as follows: The Russia team scored 37 points (with a total team score of 184 points), the China team scored 42 points (with a total team score of 181 points) , the South Korea team scored 36 points (with a total team score of 168 points), the Vietnam team scored 34 points (with a total team score of 168 points), and the United States team scored 26 points (with a total team score of 155 points).

The gold medal cutoff for this IMO was set at 29 points (with 39 contestants earning gold medals), the silver medal cutoff was 21 points (with 83 contestants earning silver medals), and the bronze medal cutoff was 14 points (with 131 contestants earning bronze medals).

In this IMO, no contestant achieved a perfect score of 42 points.

Problem 2.18 (IMO 53-6, proposed by Serbia). Find all positive integers n for which there exist non-negative integers $a_1, a_2, \ldots, a_n$ such that

$$\frac{1}{2^{a_1}} + \frac{1}{2^{a_2}} + \cdots + \frac{1}{2^{a_n}} = \frac{1}{3^{a_1}} + \frac{2}{3^{a_2}} + \cdots + \frac{n}{3^{a_n}} = 1. \tag{1}$$

Solution. Suppose that n satisfies condition (1), that is, there exist non-negative integers $a_1 \leq a_2 \leq \cdots \leq a_n$, such that $\sum_{i=1}^{n} 2^{-a_i} = \sum_{i=1}^{n} i \cdot 3^{-a_i} = 1$.

Multiplying both sides of the second equality by a_n, and then taking mod 2, we have

$$\sum_{i=1}^{n} i = \frac{1}{2} n(n+1) \equiv 1 \pmod 2.$$

Hence, $n \equiv 1, 2 \pmod 4$. In the following, we show that this is also a sufficient condition. That is, the sought n are all integers such that $n \equiv 1, 2 \pmod 4$.

We call a set of finite repeatable positive integers $B = \{b_1, b_2, \ldots, b_n\}$ as "feasible", if the elements of B can correspond to non-negative integers $a_1, a_2, \ldots, a_n$ such that

$$\sum_{i=1}^{n} 2^{-a_i} = \sum_{i=1}^{n} b_i 3^{-a_i} = 1.$$

Note an important fact that if B is "feasible", then by replacing any element b of B by two positive integers u and v with $u + v = 3b$, then the resulting set B' is also "feasible". In fact, if b corresponds to non-negative integer a, then we take both u and v corresponding to $a + 1$, keeping others' correspondence unchanged. Since

$$2^{-a-1} + 2^{-a-1} = 2^{-a}, u \cdot 3^{-a-1} + v \cdot 3^{-a-1} = b \cdot 3^{-a},$$

the set B' is "feasible". If B' can be obtained by such finite replacements from B, we denote them by $B \mapsto B'$. Particularly, if $b \in B$, then we can replace b by b and $2b$, thus, $B \mapsto B \bigcup \{2b\}$.

We shall show that for each positive integer $n \equiv 1, 2 \pmod 4$, the set $B_n = \{1, 2, \ldots, n\}$ is "feasible". The set B_1 is "feasible", since we can take $a_1 = 0$. The set B_2 is "feasible", since $B_1 \mapsto B_2$. If B_n is "feasible" for $n \equiv 1 \pmod 4$, then since $B_n \mapsto B_{n+1}$, we see that B_{n+1} is also "feasible".

Thus, it suffices to show that B_n is "feasible" for $n \equiv 1 \pmod 4$. The set B_5 is "feasible", since

$$B_2 \mapsto \{1, 3, 3\} \mapsto \{1, 3, 4, 5\} \mapsto B_5.$$

The set B_9 is "feasible", since

$$B_5 \mapsto \{1, 2, 3, 4, 6, 9\} \mapsto \{1, 2, 3, 4, 5, 6, 7, 9\} \mapsto B_9 \backslash \{8\} \mapsto B_9.$$

The set B_{13} is "feasible", since

$$B_9 \mapsto \{1, 2, 3, 4, 5, 6, 7, 9, 11, 13\} \mapsto B_{13}.$$

The last step is obtained by appending $2b$ several times. Appending 8,10, and 12 successively, we get that B_{17} is "feasible", since

$$B_6 \mapsto B_5 \bigcup \{7, 11\} \mapsto B_8 \bigcup \{11\} \mapsto B_7 \bigcup \{9, 11, 15\}$$

$$\mapsto B_{12} \bigcup \{15\} \mapsto B_{17} \backslash \{10, 14, 16\} \mapsto B_{17}.$$

Lastly, we show that $B_{4k+2} \mapsto B_{4k+13}$ for any integer $k \geq 2$, and complete the proof. By successively appending $4k + 4, 4k + 6, \ldots, 4k + 12$, we note that

$$\frac{4k + 12}{2} \leq 4k + 2.$$

In the remaining six odd numbers $4k + 3, 4k + 5, \ldots, 4k + 13$, denote the numbers that are multiples of 3 by u_1, v_1, the sum of two of which is a multiple of 3 by u_2, v_2 and u_3, v_3. Then substitute $b_i = \frac{1}{3}(u_i + v_i)$ by u_i, v_i for $i = 1, 2, 3$. Note that b_i is even, so we append $\frac{b_i}{2}$ and get B_{4k+13}.

【Score Situation】 This particular problem saw the following distribution of scores among contestants: 10 contestants scored 7 points, 1 contestant scored 6 points, no contestant scored 5 points, 9 contestants scored 4 points, 3 contestants scored 3 points, 12 contestants scored 2 points, 39 contestants scored 1 point, and 473 contestants scored 0 point. The average score of this problem is 0.336, indicating that it was extremely difficult.

Among the top five teams in the team scores, the scores of this problem are as follows: The South Korea team scored 23 points (with a total team score of 209 points), the China team scored 30 points (with a total team score of 195 points), the United States team scored 18 points (with a total team score of 194 points), the Russia team scored 9 points (with a total team score of 177 points) , the Canada team scored 13 points (with a total team score of 159 points), and the Thailand team scored 2 points (with a total team score of 159 points).

The gold medal cutoff for this IMO was set at 28 points (with 51 contestants earning gold medals), the silver medal cutoff was 21 points (with 88 contestants earning silver medals), and the bronze medal cutoff was 14 points (with 137 contestants earning bronze medals).

In this IMO, only one contestant achieved a perfect score of 42 points, namely Jeck Lim from Singapore.

2.2.3 *Exploring relationships between terms*

Problem 2.19 (IMO 13-3, proposed by Poland). Prove that the set of integers of the form $2^k - 3(k = 2, 3, \ldots)$ contains an infinite subset in which every two members are relatively prime.

Proof 1. We prove by induction. First we can see that $2^2 - 3 = 1$, $2^3 - 3 = 5$, and $2^4 - 3 = 13$ are coprime.

Now assume that there exists a subsequence $\{a_k\}$ of $\{2^n - 3\}$ such that $a_1 = 2^{n_1} - 3, a_2 = 2^{n_2} - 3, \ldots, a_k = 2^{n_k} - 3$ are coprime. We shall prove that there exists a positive integer n_{k+1} such that $a_{k+1} = 2^{n_{k+1}} - 3$ is relatively prime to every term of $\{a_k\}$. Then by mathematical induction the subset we sought exists.

To show this, we suppose $\{p_1, p_2, \ldots, p_m\}$ is the set of all the prime factors of terms of $\{a_k\}$. Obviously $p_1, p_2, \ldots, p_m$ are odd. Thus, by Fermat's little theorem, for $i = 1, 2, \ldots, m$,

$$2^{p_i - 1} \equiv 1 \pmod{p_i}.$$

Hence, for $i = 1, 2, \ldots, m$,

$$2^{(p_1 - 1)(p_2 - 1) \cdots (p_m - 1)} \equiv 1 \pmod{p_i}.$$

So we let $n_{k+1} = (p_1 - 1)(p_2 - 1) \cdots (p_m - 1)$. Then for $i = 1, 2, \ldots, m$

$$2^{n_{k+1}} - 3 \equiv 1 - 3 \equiv -2 \pmod{p_i}.$$

Hence $2^{n_{k+1}} - 3$ is not divisible by p_i. Therefore, $2^{n_{k+1}} - 3$ is relatively prime to $a_1, a_2, \ldots, a_k$. We can choose n_{k+1} to be the next term of the subsequence. By mathematical induction, the original proposition holds.

Proof 2. Like the first proof, we also prove by induction that there exists a positive integer n_{k+1} such that $a_{k+1} = 2^{n_{k+1}} - 3$ is relatively prime to every term of $\{a_k\}$, a subsequence of $\{2^n - 3\}$, whose terms $a_1 = 2^{n_1} - 3, a_2 = 2^{n_2} - 3, \ldots, a_k = 2^{n_k} - 3$ are coprime.

Now let $l = (2^{n_1} - 3)(2^{n_2} - 3) \cdots (2^{n_k} - 3)$. Then l is odd. For $2^0, 2^1, \ldots, 2^l$, by the pigeonhole principle, there exist two distinct integer s and t with $0 \leq s < t \leq l$, such that

$$2^s \equiv 2^t \pmod{l},$$

i.e.,

$$l \mid 2^s (2^{t-s} - 1).$$

For odd l, it must be $l \mid 2^{t-s} - 1$. Since $2^{t-s} - 1$ is coprime to $2^{t-s} - 3$, then let $n_{k+1} = t - s$. We see that l is coprime to $2^{n_{k+1}} - 3$, thus completing the proof.

Remark. In order to find the infinite subset, it is a natural idea to gradually expand our existing subsequence by mathematical induction. It is worth mentioning that these two proofs are essentially the same, because the steps we constructed n_{k+1} in the second proof happen to be one of the common methods for proving Fermat's little theorem which we used in the first proof.

【Score Situation】This particular problem saw the following distribution of scores among contestants: 2 contestants scored 9 points, no contestant scored 8 points, no contestant scored 7 points, no contestant scored 6 points, no contestant scored 5 points, no contestant scored 4 points, no contestant scored 3 points, no contestant scored 2 points, 4 contestants scored 1 point, and 22 contestants scored 0 point. The average score of this problem is 0.786, indicating that it was extremely difficulty.

Among the top five teams in the team scores, the scores of this problem are as follows: The Hungary team scored 45 points (with a total team score of 255 points), the Soviet Union team scored 27 points (with a total team score of 205 points), the German Democratic Republic team scored 30 points (with a total team score of 142 points), the Poland team scored 20 points (with a total team score of 118 points) , the United Kingdom team scored 23 points (with a total team score of 110 points), and the Romania team scored 20 points (with a total team score of 110 points).

The gold medal cutoff for this IMO was set at 35 points (with 7 contestants earning gold medals), the silver medal cutoff was 23 points (with 12 contestants earning silver medals), and the bronze medal cutoff was 11 points (with 29 contestants earning bronze medals).

In this IMO, only one contestant achieved a perfect score of 42 points, namely Imre Ruzsa from Hungary.

Problem 2.20 (IMO 26-2, proposed by Australia). Let n and k be given relatively prime natural numbers with $k < n$. Each number in the set $M = \{1, 2, \ldots, n-1\}$ is colored either blue or white. It is given that:

(i) for each $i \in M$, both i and $n - i$ have the same color;
(ii) for each $i \in M$ with $i \neq k$, both i and $|k - i|$ have the same color.

Prove that all numbers in M must have the same color.

Proof. For $m = 1, 2, \ldots, n-1$, suppose that mk divided by n gives q_m as the quotient and r_m as the remainder, i.e.,

$$mk = nq_m + r_m.$$

Since n is relatively prime to k, there is $1 \leq r_m \leq n-1$.

If there exists $r_m = r_{m'}$, then $(m - m')k = (q_m - q_{m'})n$. As n is relatively prime to k, then $n \mid m - m'$. However, $-n < m - m' < n$, so it must be $m = m'$. This implies that $r_1, r_2, \ldots, r_{n-1}$ are different from each other. Hence the set $\{r_1, r_2, \ldots, r_{n-1}\}$ is actually $\{1, 2, \ldots, n-1\}$, which is the given M itself.

Note that $r_{n-1} = n - k$, so for $m = 1, 2, \ldots, n-2$, we have $r_m \neq n - k$. Thus we can discuss the following two cases:

Case 1: $r_m < n - k$. Then $r_{m+1} = r_m + k$. By condition (ii) we know that r_{m+1} and r_m have the same color.

Case 2: $r_m > n - k$. Then $r_{m+1} = r_m + k - n$. By condition (ii) we know r_{m+1} and $k - r_{m+1} = n - r_m$ have the same color; next, by condition (i) we know $n - r_m$ and r_m have the same color, which also leads to that r_{m+1} and r_m have the same color.

Summarizing, we see that for $m = 1, 2, \ldots, n - 2$, it always holds that r_{m+1} and r_m have the same color, which implies that all numbers in M must have the same color.

Remark. The reason why this problem was classified as a number theory problem instead of combinatorics is the important number theory idea it contains: for two coprime positive integers k and n, when m goes through the residue classes modulo n, the integer mk will also go through the residue classes modulo n.

【Score Situation】 This particular problem saw the following distribution of scores among contestants: 92 contestants scored 7 points, 6 contestants scored 6 points, 4 contestant scored 5 points, 3 contestants scored 4 points, 9 contestants scored 3 points, 8 contestants scored 2 points, 27 contestants scored 1 point, and 60 contestants scored 0 point. The average score of this problem is 3.742, indicating that it was relatively straightforward.

Among the top five teams in the team scores, the scores of this problem are as follows: The Romania team scored 42 points (with a total team score of 201 points), the United States team scored 38 points (with a total team score of 180 points), the Hungary team scored 31 points (with a total team score of 168 points), the Bulgaria team scored 35 points (with a total team score of 165 points), and the Vietnam team scored 35 points (with a total team score of 144 points).

The gold medal cutoff for this IMO was set at 34 points (with 14 contestants earning gold medals), the silver medal cutoff was 22 points (with 35 contestants earning silver medals), and the bronze medal cutoff was 15 points (with 52 contestants earning bronze medals).

In this IMO, only two contestants achieved a perfect score of 42 points, namely Géza Kós from Hungary and Daniel Tătaru from Romania.

Problem 2.21 (IMO 42-4, proposed by Canada). Let n be an odd integer greater than 1, and let $k_1, k_2, \ldots, k_n$ be given integers. For each of the $n!$ permutations $a = (a_1, a_2, \ldots, a_n)$ of $1, 2, \ldots, n$, let

$$S(a) = \sum_{i=1}^{n} k_i a_i.$$

Prove that there are two permutations b and c with $b \neq c$, such that $n!$ is a divisor of $S(b) - S(c)$.

Proof. We prove it by contradiction.

Assume that for any two different permutations b and c,

$$S(b) \not\equiv S(c) \ (\mathrm{mod} \ n!).$$

This leads to that when a goes through the $n!$ permutations of $1, 2, \ldots, n$, the number $S(a)$ will go through the residue classes modulo $n!$ correspondingly (each residue class once). Then

$$\sum_a S(a) \equiv 1 + 2 + \cdots + n!$$

$$= \frac{n!}{2}(n! + 1)$$

$$\equiv \frac{n!}{2} \ (\mathrm{mod} \ n!), \tag{1}$$

where $\sum_a$ means the sum is for a going through the $n!$ permutations of $1, 2, \ldots, n$.

On the other hand,

$$\sum_a S(a) = \sum_a \sum_{i=1}^{n} k_i a_i$$

$$= \sum_{i=1}^{n} k_i \sum_a a_i$$

$$= \sum_{i=1}^{n} k_i \cdot \frac{n(n+1)}{2} \cdot (n-1)!$$

$$= n! \left(\frac{n+1}{2} \sum_{i=1}^{n} k_i \right). \tag{2}$$

Since n is an odd integer greater than 1, by (2) we have $\sum_a S(a) \equiv 0 (\mathrm{mod} \ n!)$, which is a contradiction to (1). The proof is completed.

Remark. Here the problem requires that $S(a)$ cannot go through the residue classes modulo $n!$, so we came up with the idea of proving it by contradiction, in order to take advantage of the property that $S(a)$ go through the residue classes. Then we used the "double counting" method, which is common in combinatorics problems.

【Score Situation】 This particular problem saw the following distribution of scores among contestants: 173 contestants scored 7 points, 8 contestants scored 6 points, 6 contestants scored 5 points, 15 contestants scored 4 points, 11 contestants scored 3 points, 34 contestants scored 2 points, 79 contestants scored 1 point, and 147 contestants scored 0 point. The average score of this problem is 3.233, indicating that it was relatively straightforward.

Among the top five teams in the team scores, the scores of this problem are as follows: The China team scored 42 points (with a total team score of 225 points), the Russia team scored 42 points (with a total team score of 196 points), the United States team scored 42 points (with a total team score of 196 points) , the Bulgaria team scored 42 points (with a total team score of 185 points), and the South Korea team scored 42 points (with a total team score of 185 points).

The gold medal cutoff for this IMO was set at 30 points (with 39 contestants earning gold medals), the silver medal cutoff was 20 points (with 81 contestants earning silver medals), and the bronze medal cutoff was 11 points (with 122 contestants earning bronze medals).

In this IMO, a total of four contestants achieved a perfect score of 42 points.

Problem 2.22 (IMO 46-2, proposed by the Netherlands).

Let $a_1, a_2, \ldots$ be a sequence of integers with infinitely many positive and negative terms. Suppose that for every positive integer n the numbers $a_1, a_2, \ldots, a_n$ leave n different remainders upon division by n.

Prove that every integer occurs exactly once in the sequence $a_1, a_2, \ldots$.

Proof. First we will show that every positive integer will occur at most once in the sequence. Note that if $a_i = a_j = k \, (i < j)$, then two numbers among $a_1, a_2, \ldots, a_j$, say a_i, a_j, are congruent modulo j, which is impossible.

Let x_k and y_k be the greatest and the smallest number among $a_1, a_2, \ldots, a_k$ respectively. Then $x_k - y_k \leq k - 1$. For if $x_k - y_k \geq k$, without loss of generality, let $a_i = x_k, a_j = y_k$, and $a_j - a_i = l \geq k$.

Then $i, j \leq k \leq l$. Therefore two numbers among $a_1, a_2, \ldots, a_l$, say a_i, a_j, are congruent modulo l, which is impossible.

Now we will show that for every integer $t(y_k \leq t \leq x_k)$, there exists an integer $s(1 \leq s \leq k)$ such that $a_s = t$. For, if $a_1, a_2, \ldots, a_k \in \{u \in \mathbf{Z} | y_k \leq u \leq x_k, u \neq t\}$, then the sequence has $x_k - y_k$ different values at most. Note that $x_k - y_k \leq k - 1 < k$. Therefore two numbers among $a_1, a_2, \ldots, a_k$ have the same value, which is a contradiction to the above argument.

Now for any integer m, since there are infinity many negative and positive numbers in the sequence, it is trivial to see that there exists a positive integer p such that $a_p > |m|$. By a similar argument, there exists a positive integer q such that $a_q < -|m|$. Denote $r = \max\{p, q\}$. Then $x_r > |m|, y_r < -|m|$, i. e., $y_r < m < x_r$. The above arguments lead us to conclude that there exists a positive integer s such that $a_s = m$. We conclude that every number must appear exactly once in the sequence.

Remark. If we try to write a desired sequence term by term, we will find that each newly added term cannot be the same as any existing term, nor can it be too far away from the existing terms. So we proved these two properties in "mathematical language" and then the original problem was roughly solved.

【Score Situation】 This particular problem saw the following distribution of scores among contestants: 175 contestants scored 7 points, 2 contestants scored 6 points, 7 contestants scored 5 points, 11 contestants scored 4 points, 7 contestants scored 3 points, 13 contestants scored 2 points, 202 contestants scored 1 point, and 96 contestants scored 0 point. The average score of this problem is 3.051, indicating that it was relatively straightforward.

Among the top five teams in the team scores, the scores of this problem are as follows: The China team scored 42 points (with a total team score of 235 points), the United States team scored 36 points (with a total team score of 213 points), the Russia team scored 42 points (with a total team score of 212 points), the Iran team scored 42 points (with a total team score of 201 points), and the South Korea team scored 42 points (with a total team score of 200 points).

The gold medal cutoff for this IMO was set at 35 points (with 42 contestants earning gold medals), the silver medal cutoff was 23 points (with 79 contestants earning silver medals), and the bronze medal cutoff was 12 points (with 128 contestants earning bronze medals).

In this IMO, a total of 16 contestants achieved a perfect score of 42 points.

2.2.4 *Maximum or minimum value problems*

Problem 2.23 (IMO 20-1, proposed by Cuba). Two natural numbers m and n satisfy $1 \le m < n$. In their decimal representations, the last three digits of 1978^m are equal, respectively, to the last three digits of 1978^n. Find m and n such that $m + n$ has its least value.

Solution. The given condition is equivalent to

$$1978^n - 1978^m = 1978^m(1978^{n-m} - 1) \equiv 0(\bmod\ 10^3).$$

Since $(2, 1978^{n-m} - 1) = 1$, $(5, 1978^m) = 1$, then

$$1978^m \equiv 0(\bmod\ 8), \tag{1}$$
$$1978^{n-m} \equiv 1(\bmod\ 125). \tag{2}$$

By equation (1) we know $m \ge 3$. Note that $1978 \equiv 3(\bmod\ 5)$. Then by equation (2), $3^{n-m} \equiv 1(\bmod\ 5)$, and hence $4|n-m$. Let $n-m = 4k, k \in \mathbf{N}^*$. Then

$$1978^{n-m} \equiv (-22)^{4k} \equiv (-16)^{2k} \equiv 6^k \equiv 1(\bmod\ 125).$$

We can find that $6^k - 1 = (5+1)^k - 1 \equiv 5^2 \cdot \frac{k(k-1)}{2} + 5k(\bmod\ 125)$. Thus

$$25|5 \cdot \frac{k(k-1)}{2} + k,$$

i.e.,

$$25|\frac{k(5k-3)}{2}.$$

Since $(25, 5k-3) = 1$, so $25|k$. Hence $100|n-m$, which implies $n-m \ge 100$.

Summarizing, we find that $m + n = 2m + (n-m) \ge 2 \times 3 + 100 = 106$.

For $m = 3$, we have $n = 103$, and by the congruence equations above we know this pair of m and n satisfies the given condition, so the least value we want is 106.

Remark. In this solution, we can also use the fact that the smallest $n - m$ must be a factor of $\varphi(125) = 100$. Then combined with $4|n - m$ we know $n-m$ must be one of 4, 20, or 100. Of course, there are many other methods discussing congruence modulo 125, and it is even feasible to find the order of 1978 modulo 5, 25, and 125 by enumeration.

【Score Situation】 This particular problem saw the following distribution of scores among contestants: 27 contestant scored 6 points, 4 contestants scored 5 points, 2 contestants scored 4 points, 3 contestants scored 3 points, 4 contestants scored 2 points, 6 contestants scored 1 point, and 2 contestants scored 0 point. The average score of this problem is 4.438, indicating that it was simple.

Among the top five teams in the team scores, the scores of this problem are as follows: The Romania team scored 46 points (with a total team score of 237 points), the United States team scored 44 points (with a total team score of 225 points), the United Kingdom team scored 43 points (with a total team score of 201 points), the Vietnam team scored 45 points (with a total team score of 200 points), and the Czechoslovakia team scored 41 points (with a total team score of 195 points).

The gold medal cutoff for this IMO was set at 35 points (with 5 contestants earning gold medals), the silver medal cutoff was 27 points (with 20 contestants earning silver medals), and the bronze medal cutoff was 22 points (with 38 contestants earning bronze medals).

In this IMO, only one contestant achieved a perfect score of 40 points, namely Mark Kleiman from the United States.

Problem 2.24 (IMO 37-4, proposed by Russia). Two positive integers a and b are such that the numbers $15a + 16b$ and $6a - 15b$ are both squares of positive integers. What is the least possible value that can be taken on by the smaller of these two squares?

Solution. Suppose positive integers x and y satisfy

$$15a + 16b = x^2, 16a - 15b = y^2.$$

By squaring these two equalities and adding them up, we have

$$x^4 + y^4 = (15^2 + 16^2)(a^2 + b^2) = 481(a^2 + b^2).$$

Thus, $481 | x^4 + y^4$. We will show that both x and y are divisible by $481 = 13 \times 37$.

Assume for the contrary that $13 \nmid x$. Since $x^4 + y^4 \equiv 0 \pmod{13}$, it holds that $13 \nmid y$. By Fermat's little theorem,

$$1 \equiv x^{12} \equiv (x^4)^3 \equiv (-y^4)^3 \equiv -y^{12} \equiv -1 \pmod{13},$$

a contradiction. So $13 | x$ and $13 | y$.

Similarly, assume that $37 \nmid x$. Since $x^4 + y^4 \equiv 0 \pmod{37}$, it holds that $37 \nmid y$. By Fermat's little theorem,

$$1 \equiv x^{36} \equiv (x^4)^9 \equiv (-y^4)^9 \equiv -y^{36} \equiv -1 \pmod{37},$$

also a contradiction. So $37|x$ and $37|y$.

Therefore, we conclude that $481|x$, $481|y$, and $x, y \geq 481$.

On the other hand, take $a = 481 \times 31$ and $b = 481$. Then $15a + 16b = 16a - 15b = 481^2$. Thus the minimum value we looked for is $481^2 = 231361$.

Remark. Besides this solution, there are other ways to prove that $481|x$ and $481|y$. These methods are similar so we will not list them. As we can see, Fermat's little theorem is quite useful when discussing an integer modulo a prime number. In fact, for any odd prime number p, let $p - 1 = st$, where t is an odd number. Then like what we have done in this solution, we can prove $x^s + y^s \equiv 0 \pmod{p} \Leftrightarrow p|x, p|y$ by Fermat's little theorem.

【Score Situation】This particular problem saw the following distribution of scores among contestants: 86 contestants scored 7 points, 6 contestants scored 6 points, 7 contestants scored 5 points, 11 contestants scored 4 points, 17 contestants scored 3 points, 11 contestants scored 2 points, 109 contestants scored 1 point, and 177 contestants scored 0 point. The average score of this problem is 2.120, indicating that it had a certain level of difficulty.

Among the top five teams in the team scores, the scores of this problem are as follows: The Romania team scored 42 points (with a total team score of 187 points), the United States team scored 35 points (with a total team score of 185 points), the Hungary team scored 36 points (with a total team score of 167 points), the Russia team scored 36 points (with a total team score of 162 points), and the United Kingdom team scored 39 points (with a total team score of 161 points).

The gold medal cutoff for this IMO was set at 28 points (with 35 contestants earning gold medals), the silver medal cutoff was 20 points (with 66 contestants earning silver medals), and the bronze medal cutoff was 12 points (with 99 contestants earning bronze medals).

In this IMO, only one contestant achieved a perfect score of 42 points, namely Ciprian Manolescu from Romania.

Problem 2.25 (IMO 57-4, proposed by Luxembourg). A set of positive integers is called fragrant if it contains at least two elements and each of its elements has a prime factor in common with at least one of the other

elements. Let $P(n) = n^2 + n + 1$. What is the least possible positive integer value of b such that there exists a non-negative integer a for which the set

$$\{P(a+1), P(a+2), \ldots, P(a+b)\}$$

is fragrant?

Solution. The minimum value of b is 6.

We first prove some facts.

(i) $(P(n), P(n+1)) = 1$.

This is because

$$\begin{aligned}(P(n), P(n+1)) &= (n^2 + n + 1, (n+1)^2 + (n+1) + 1)\\ &= (n^2 + n + 1, 2n + 2)\\ &= (n^2 + n + 1, n + 1) = 1.\end{aligned}$$

(ii) $(P(n), P(n+2)) = 7$ when $n \equiv 2 \pmod 7$; otherwise, $(P(n), P(n+2)) = 1$.

Observe that

$$\begin{aligned}(P(n), P(n+2)) &= (n^2 + n + 1, 4n + 6)\\ &= (n^2 + n + 1, 2n + 3)\\ &= (4n^2 + 4n + 4, 2n + 3)\\ &= (7, 2n + 3).\end{aligned}$$

Only when $7 | 2n + 3$, i.e., $n \equiv 2 \pmod 7$, we have $(P(n), P(n+2)) = 7$; otherwise, $(P(n), P(n+2)) = 1$.

(iii) $(P(n), P(n+3)) = 3$ when $n \equiv 1 \pmod 3$; otherwise, $(P(n), P(n+3)) = 1$.

By checking modulo 9, when $n \equiv 1 \pmod 3$, we have $3 | n^2 + n + 1$, $9 \nmid n^2 + n + 1$, and $9 | 6n + 12$. So,

$$\begin{aligned}(P(n), P(n+3)) &= (n^2 + n + 1, 6n + 12)\\ &= (n^2 + n + 1, n + 2)\\ &= (3, n + 2).\end{aligned}$$

(iv) $(P(n), P(n+4)) = 19$ when $n \equiv 7$ (mod 19); otherwise, $(P(n), P(n+4)) = 1$.

$$\begin{aligned}
(P(n), P(n+4)) &= (n^2 + n + 1, 8n + 20) \\
&= (n^2 + n + 1, 2n + 5) \\
&= (4n^2 + 4n + 4, 2n + 5) \\
&= (19, 2n + 5).
\end{aligned}$$

Only when $19 | 2n + 5$, i.e., $n \equiv 7$ (mod 19), we have $(P(n), P(n+4)) = 19$; otherwise,

$$(P(n), P(n+4)) = 1.$$

Now we prove that there are no required fragrant sets for $b \leq 5$.

When $b = 2$, we see that $P(a+1)$ and $P(a+2)$ are coprime for any a.

When $b = 3$, it holds that $P(a+2)$ is coprime to both $P(a+1)$ and $P(a+3)$.

When $b = 4$, suppose such a fragrant set exists. Then $P(a+2)$ and $P(a+4)$ are not coprime, and so are $P(a+1)$ and $P(a+3)$. This implies $a+1 \equiv a+2 \equiv 2$ (mod 7), which is impossible.

When $b = 5$, and suppose such a fragrant set exists. Then $P(a+3)$ is not coprime to at least one of $P(a+1)$ and $P(a+5)$. It follows that $a+1 \equiv 2$ (mod 7) or $a+3 \equiv 2$ (mod 7). Now $P(a+2)$ and $P(a+4)$ are coprime, so we must have $P(P(a+1), P(a+4)) > 1$ and $P(P(a+2), P(a+5)) > 1$. Hence $a+1 \equiv a+2 \equiv 1$ (mod 3), which is impossible.

Finally we show that there is a required fragrant set when $b = 6$. By the Chinese remainder theorem, there is an integer a such that

$a+1 \equiv 1$ (mod 3), $a+2 \equiv 7$ (mod 19), and $a+3 \equiv 2$ (mod 7).

Then $P(P(a+1), P(a+4)) = 3$, $P(P(a+4), P(a+6)) = 19$ and $P(P(a+3), P(a+5)) = 7$. Therefore, $\{P(a+1), P(a+2), \ldots, P(a+6)\}$ is fragrant.

Remark. It is sufficient to prove such a exists by the Chinese remainder theorem, but in fact we can easily solve these congruence equations and get one solution $a = 195$. After testing, it does indeed meet the requirements of the problem.

【Score Situation】 This particular problem saw the following distribution of scores among contestants: 347 contestants scored 7 points, 24 contestants scored 6 points, 15 contestants scored 5 points, 26 contestants scored 4 points, 10 contestants scored 3 points, 26 contestants scored 2 points, 22 contestants scored 1 point, and 132 contestants scored 0 point. The average score of this problem is 4.744, indicating that it was simple.

Among the top five teams in the team scores, the scores of this problem are as follows: The United States team scored 42 points (with a total team score of 214 points), the South Korea team scored 42 points (with a total team score of 207 points), the China team scored 42 points (with a total team score of 204 points), the Singapore team scored 39 points (with a total team score of 196 points), and the Chinese Taiwan team scored 38 points (with a total team score of 175 points).

The gold medal cutoff for this IMO was set at 29 points (with 44 contestants earning gold medals), the silver medal cutoff was 22 points (with 101 contestants earning silver medals), and the bronze medal cutoff was 16 points (with 135 contestants earning bronze medals).

In this IMO, a total of six contestants achieved a perfect score of 42 points.

2.3 Summary

The divisibility relationship of integers can be regarded as a special congruence relationship, where the remainder is zero implies being congruent to zero. Therefore, the content of this chapter is somewhat related to the discussion on divisibility in the first chapter.

In the first 64 IMOs, there were a total of 25 modular arithmetic problems. These problems can be broadly categorized into four types, as depicted in Figure 2.1. The score details for these problems are presented

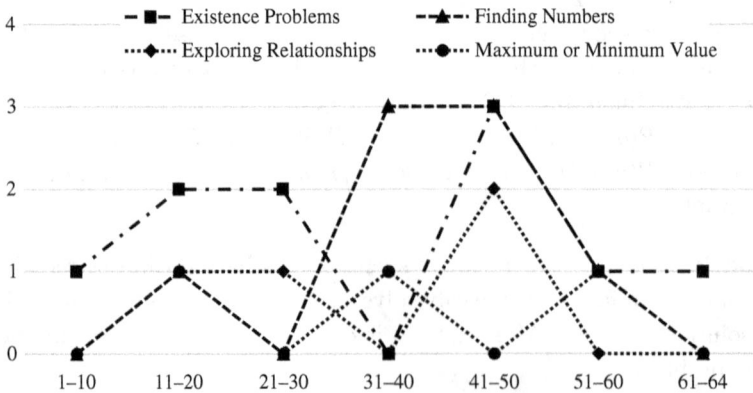

Figure 2.1 Numbers of Modular Arithmetic Problems in the First 64 IMOs

in Table 2.2. Due to the smaller number of participating teams and missing contestant score information in early IMOs, there are several blanks in Table 2.2.

Table 2.2 Score Details of Modular Arithmetic Problems in the First 64 IMOs

Problem	2.1	2.2	2.3	2.4	2.5	2.6	2.7	2.8
Average score	5.235	2.807	4.164	1.827	3.808	1.633	0.578	0.804
Top five mean	6.000	4.100	6.375	5.767	6.967	5.333	3.167	4.567

Problem	2.9	2.10	2.11	2.12	2.13	2.14	2.15	2.16
Average score	0.294	0.808	3.492	2.091	1.058	2.809	1.257	3.758
Top five mean	3.000	4.000	4.925	4.300	4.000	6.033	5.300	7.000

Problem	2.17	2.18	2.19	2.20	2.21	2.22	2.23	2.24
Average score	1.898	0.336	0.786	3.742	3.233	3.051	4.438	2.120
Top five mean	5.833	2.639	3.438	6.033	7.000	6.800	5.475	6.267

Problem	2.25
Average score	4.744
Top five mean	6.767

Note. Top five mean = Total score of the top five teams ÷ Total number of contestants from the top five teams.

Problem 2.1–2.10 focus on "existence problems"; Problems 2.11–2.18 deal with "finding numbers that satisfy given conditions"; Problems 2.19–2.22 are about "exploring relationships between terms"; Problems 2.23–2.25 concern "maximum or minimum value problems".

These 25 problems were proposed by 18 countries. Poland contributed three problems; Romania, the United Kingdom, the Soviet Union, Russia, and the United States each contributed two problems.

From Table 2.2, it can be observed that in the first 64 IMOs, there were six modular arithmetic problems with an average score of 0–1 points; five problems with an average score of 1–2 points; four problems with an average score of 2–3 points; six problems with an average score of 3–4 points; four problems with an average score above 4 points. Overall, the modular arithmetic problems were relatively challenging. Problems with an average score below 2 points accounted for approximately half of all modular arithmetic problems.

Among these 25 problems, the problems with the lowest average score are Problem 2.7 (IMO 44-6, contributed by France, existence problems),

Problem 2.8 (IMO 49-3, contributed by Lithuania, existence problems), Problem 2.9 (IMO 58-6, contributed by the United States, existence problems), Problem 2.10 (IMO 63-3, contributed by the United States, existence problems), Problem 2.18 (IMO 53-6, contributed by Serbia, finding numbers that satisfy given conditions), and Problem 2.19 (IMO 13-3, contributed by Poland, exploring relationships between terms). There are four existence problems here, which were very difficult.

Chapter 3

Indeterminate Equations

Indeterminate equations may have different definitions. Generally speaking, equations or systems of equations with more unknowns than equations can be referred to as indeterminate equations. Typically, the indeterminate equations we discuss and try to solve are those with integer coefficients; and only integer solutions are of interest.

As early as in the 3rd century AD, ancient Greece mathematician Diophantus conducted research on solving some indeterminate equations, restricted to integer solutions. Therefore, some indeterminate equations may also be called Diophantine equations. The most basic type of indeterminate equations, or say Diophantine equations, are those polynomial equations with integer coefficients, where the goal is to find integer solutions. Building upon this foundation, there are equations that involve unknowns in the exponents, which are also included in the discussion of indeterminate equations. For example, the equations in the famous Fermat's last theorem (stating that for integer n greater than 2, the equation $x^n + y^n = z^n$ has no integer solutions) falls into this category.

In the first 64 IMOs, there had been a total of 14 indeterminate equation problems, accounting for approximately 18.7% of all number theory problems. These problems can be primarily categorized into two types: (1) finding solutions of indeterminate equations, totaling 10 problems; (2) proving properties satisfied by indeterminate equations, totaling four problems. The statistical distribution of these two types of problems in previous IMOs is presented in Table 3.1.

It can be observed that, compared to divisibility and congruence related problems, the number of problems related to indeterminate equations is

Table 3.1 Numbers of Indeterminate Equation Problems in the First 64 IMOs

Content	Session							Total
	1–10	11–20	21–30	31–40	41–50	51–60	61–64	
Finding solutions	2	1	1	1	2	2	1	10
Proving properties	0	0	3	0	0	1	0	4
Number theory problems	7	10	12	16	14	11	5	74
The percentage of indeterminate equation problems among number theory problems	28.6%	10.0%	33.3%	6.3%	14.3%	27.3%	20.0%	18.7%

relatively small. The highest proportion occurred in the 11th to 20th IMO, accounting for 33.3%.

In fact, finding solutions of indeterminate equations and proving properties satisfied by indeterminate equations both require knowledge and methods related to divisibility and congruence, such as prime factor analysis. Therefore, we can effectively assess students' grasp of foundational number theory knowledge through indeterminate equation problems.

This chapter will be divided into three parts. The first part introduces some common indeterminate equations, such as Pythagorean equations and Pell's equations, the latter of which is sometimes associated with the method of infinite descent. Then common methods for solving indeterminate equations are provided.

The second part revolves around two types of problems: "finding solutions of indeterminate equations" and "proving properties satisfied by indeterminate equations". These problems are presented in chronological order, and some problems include various solutions and generalizations.

It is important to note that for each problem, the solutions are followed by information on the scores, including the number of contestants in each score range, the average score, and the scores of the top five teams. However, early IMOs often lacked information on contestant scores, so the number of contestants in each score range only represents the counted number of contestants, and some problems lack scores of the top five teams.

The third part provides a brief summary of this chapter.

3.1 Common Properties, Theorems, and Methods

3.1.1 *Common indeterminate equations*

(1) *Linear equations in two variables*

Theorem 3.1. *If a, b are coprime integers, c is an integer, and equation*

$$ax + by = c \tag{1}$$

has a pair of integer solutions x_0, y_0, then all pairs of integer solutions of this equation are

$$\begin{cases} x = x_0 - bt, \\ y = y_0 + at, \end{cases}$$

where $t = 0, \pm 1, \pm 2, \pm 3, \ldots$.

Proof. Since x_0, y_0 form a solution to equation (1), they satisfy

$$ax_0 + by_0 = c. \tag{2}$$

Therefore,

$$a(x_0 - bt) + b(y_0 + at) = ax_0 + by_0 = c.$$

This implies that $x = x_0 - bt$, $y = y_0 + at$ is also a solution to equation (1).

On the other hand, suppose x', y' is a solution to equation (1). Then

$$ax' + by' = c. \tag{3}$$

(3)–(2) will lead to

$$a(x' - x_0) = -b(y' - y_0). \tag{4}$$

Since $(a, b) = 1$, then $a | y' - y_0$, i.e., $y' = y_0 + at$, where t is an integer. By substituting $y' = y_0 + at$ into (4), we have $x' = x_0 - bt$. Hence x', y' can be written as $x = x_0 - bt$, $y = y_0 + at$. As a conclusion, $x = x_0 - bt$, $y = y_0 + at$ denote all the solutions of equation (1).

With the theorem 1, we know that the key to solving linear equations in two variables is to find one particular solution.

Example 3.1. Find the integer solutions to equation $11x + 15y = 7$.

Solution 1. We transform the equation into the form

$$x = \frac{7 - 15y}{11}.$$

Since x is an integer, $7 - 15y$ is divisible by 11. We observe that $x_0 = 2$, $y_0 = -1$ is a solution to the equation. Therefore, the integer solutions to the original equation are

$$\begin{cases} x = 2 - 15t, \\ y = -1 + 11t, \end{cases}$$

where t is an integer.

Solution 2. First, we consider the equation $11x + 15y = 1$. Observing

$$11 \times (-4) + 15 \times (3) = 1,$$

we have

$$11 \times (-4 \times 7) + 15 \times (3 \times 7) = 7.$$

Thus we can choose $x_0 = -28$, $y_0 = 21$. Therefore,

$$\begin{cases} x = -28 - 15t, \\ y = 21 + 11t, \end{cases}$$

where t is an integer.

It can be seen that in the absence of constraints, linear indeterminate equations in two variables usually have infinitely many integer solutions. Due to the different particular solutions we choose, the form of the solutions to the same indeterminate equations can be different. However, all the solutions in different forms are actually the same. By appropriately substituting the parameter t in the solution, the solutions can be transformed into the same form.

Linear equations in more than two variables can be transformed into linear equations in two variables.

Example 3.2. Find the integer solutions to the equation $9x + 24y - 5z = 1000$.

Solution. Suppose $9x + 24y = 3t$, i.e., $3x + 8y = t$. Then $3t - 5z = 1000$. Now the original equation is equivalent to

$$\begin{cases} 3x + 8y = t, & (1) \\ 3t - 5z = 1000. & (2) \end{cases}$$

Similar to the previous method, we can find the solutions to (1) as

$$\begin{cases} x = 3t - 8u, \\ y = -t + 3u, \end{cases} \quad u \text{ is an integer.}$$

The solutions to (2) are

$$\begin{cases} t = 2000 + 5v, \\ z = 1000 + 3v, \end{cases} \quad v \text{ is an integer.}$$

By eliminating t, we have

$$\begin{cases} x = 6000 - 8u + 15v, \\ y = -2000 + 3u - 5v, \quad u, v \text{ are integers.} \\ z = 1000 + 3v, \end{cases}$$

(2) Pythagorean equation

A solution (x, y, z) to the indeterminate equation

$$x^2 + y^2 = z^2 \tag{1}$$

is called a Pythagorean triple. The Pythagorean triple in which x, y, z are coprime is called a primitive Pythagorean triple, and (x, y, z) is called a primitive solution to (1).

Theorem 3.2. *All the solutions (x, y, z) to the indeterminate equation*

$$x^2 + y^2 = z^2$$

which satisfy $(x, z) = 1, 2 | y$ can be represented as

$$x = a^2 - b^2, y = 2ab, z = a^2 + b^2,$$

where a, b are coprime integers and have opposite parity, with $a > b$.

Example 3.3. Prove that the indeterminate equation $x^4 + y^4 = z^2$ has no positive integer solution.

Proof. We prove it by contradiction. Assume the equation $x^4 + y^4 = z^2$ has a positive integer solution (x_0, y_0, z_0), where we suppose $(x_0, y_0) = 1$

without loss of generality, y_0 is even, and z_0 is the least z in all the solutions to the equation. Then there exist positive integers a, b, where $(a,b) = 1$, a, b have opposite parity, and $a > b$, such that

$$x_0^2 = a^2 - b^2, \quad y_0^2 = 2ab, \quad z_0 = a^2 + b^2.$$

Since x_0 is odd, then a is odd (or $x_0^2 = a^2 - b^2 \equiv 0 - 1 \equiv 3 \pmod 4$, a contradiction) and b is even. By

$$x_0^2 + b^2 = a^2$$

we know that there exist positive integers p, q, where $(p,q) = 1$, p, q have opposite parity, and $p > q$, such that

$$x_0 = p^2 - q^2, \quad b^2 = 2pq, \quad a = p^2 + q^2,$$

which implies

$$y_0^2 = 4pq(p^2 + q^2).$$

Since $(p,q) = 1$, we have $(p, p^2 + q^2) = (q, p^2 + q^2) = 1$, so $(pq, p^2 + q^2) = 1$. Therefore, there exist coprime positive integers r, s, z_1, such that

$$p = r^2, \quad q = s^2, \quad p^2 + q^2 = z_1^2.$$

Hence (r, s, z_1) is also a positive solution to the original equation. However, $z_1 < p^2 + q^2 = a < z_0$, which is a contradiction to the choice that z_0 is the least.

Summarizing, we see that there is no positive integer solution to the original equation.

(3) Pell's equation

The quadratic equation in two variables of the form

$$x^2 - Dy^2 = 1 \ (D \text{ is an integer}) \tag{1}$$

is called Pell's equation.

Suppose positive integers x_0, y_0 form a solution to equation (1) with the least $x_0 + \sqrt{D}y_0$. Then (x_0, y_0) is called the **fundamental solution** to equation (1).

Theorem 3.3. *There exist infinitely many positive integer solutions to the equation*

$$x^2 - Dy^2 = 1 \ (D \text{ is an integer}).$$

All positive integer solutions (x, y) can be denoted by the fundamental solution (x_0, y_0) in the form of

$$x + \sqrt{D}y = (x_0 + \sqrt{D}y_0)^n,$$

where n is a positive integer.

Theorem 3.4. *Let D be a positive integer, and D is not a perfect square number. If equation*

$$x^2 - Dy^2 = -1 \tag{2}$$

has a positive integer solution (x, y), then (2) has infinitely many positive integer solutions.

3.1.2 Common methods for solving indeterminate equations

(1) *Method of polynomial factorization*

For an indeterminate equation, if we can write it in the form that one side of the equation is a constant while the other side is a product of several polynomial factors, such as

$$A_1 A_2 \cdots A_m = C,$$

where $A_1, A_2, \ldots, A_m$ are algebraic expressions with unknowns and C is a constant, then we can transform the original equation into a system of equations by taking factors of C as the values of $A_1, A_2, \ldots, A_m$. This method is usually helpful for simplifying problems and is one of the fundamental methods to solve indeterminate equations.

Example 3.4. Find all positive integers n, such that the equation $\frac{1}{x} + \frac{1}{y} = \frac{1}{n}$ of x, y has exactly 2011 positive integer solutions (x, y) where $x \le y$.

Solution. First, we transform the fractional expression into the integral expression $xy - nx - ny = 0$. Then

$$(x - n)(y - n) = n^2.$$

Hence $x = y = 2n$ forms a solution. In addition, every positive factor of n^2 less than n, taken as the value of $x - n$, can lead to positive integer solutions (x, y). So the desired integers are those satisfy that the number of positive factors of n^2 less than n is exactly 2010.

Suppose the prime factorization of n is $n = p_1^{\alpha_1} p_2^{\alpha_2} \cdots p_k^{\alpha_k}$, where $p_1, p_2, \ldots, p_k$ are distinct prime numbers. Since we know that the number of positive factors for n^2 is odd, so the number of those factors less than n is

$$\frac{(2\alpha_1 + 1)(2\alpha_2 + 1) \cdots (2\alpha_k + 1) - 1}{2}.$$

Therefore,

$$\frac{(2\alpha_1 + 1)(2\alpha_2 + 1) \cdots (2\alpha_k + 1) - 1}{2} = 2010,$$

i.e.,

$$(2\alpha_1 + 1)(2\alpha_2 + 1) \cdots (2\alpha_k + 1) = 4021.$$

Note that 4021 is a prime, so $k = 1$ and $2\alpha_1 + 1 = 4021$, from which $\alpha_1 = 2010$. In summary, all the sought integers are $n = p^{2010}$, where p is a prime.

(2) Method of completing the square

As a commonly used method for algebraic problems, completing the square also plays a significant role in solving indeterminate equation problems. If one side of an indeterminate equation can be written as the sum of several squares, then it is easy to obtain integer solutions through enumeration or the congruence property of perfect squares.

Example 3.5. Find integer solutions to the equation $x^2 + x = y^4 + y^3 + y^2 + y$.

Solution. Multiplying both sides of the equation by 4 and add 1, and then converting the left hand side to the form of sums of squares, we have

$$(2x + 1)^2 = 4(y^4 + y^3 + y^2 + y) + 1.$$

Since

$$4(y^4 + y^3 + y^2 + y) + 1 = (2y^2 + y + 1)^2 - y^2 + 2y$$
$$= (2y^2 + y)^2 + 3y^2 + 4y + 1,$$

for $y < -1$ or $y > 2$, we obtain

$$(2y^2 + y)^2 < (2x + 1)^2 < (2y^2 + y + 1)^2.$$

However, $2y^2 + y$, $2y^2 + y + 1$ are two consecutive integers, so there is no perfect square between them, and thus the equation cannot hold.

For $-1 \le y \le 2$, we take $y = -1, 0, 1, 2$ separately, and get all the integer solutions to the original equation

$$(x, y) = (0, -1), (-1, -1), (0, 0), (-1, 0), (-6, 2), (5, 2).$$

(3) *Method of scaling*

For many indeterminate equations, we can use inequality scaling to determine the range of values for an unknown or an algebraic expression in the equation. Then we can use enumeration to simplify the problem and obtain the solution. The essential reason for using scaling and enumeration here is the discreteness of integers. When upper and lower bounds are limited, there are at most a finite number of integers in this range.

Example 3.6. Find non-negative integer solutions (m, n) to the equation $n(n + 2) = 4m^4 + 4m^2 - 4m + 4$.

Solution. If $n = 0$, then $4m^4 + 4m^2 - 4m + 4 = 0$. However

$$4m^4 + 4m^2 - 4m + 4 = 4\left(m^4 + \left(m - \frac{1}{2}\right)^2 + \frac{3}{4}\right) > 0,$$

which is a contradiction.

If $m = 0$, then $n^2 + 2n = 4$, which is also impossible.

Therefore, m, n are positive integers. We transform the original equation into the form

$$n^2 + 2n - 4(m^4 + m^2 - m + 1) = 0,$$

and then

$$\Delta = 4 + 16(m^4 + m^2 - m + 1) = 4(4m^4 + 4m^2 - 4m + 5)$$

is a perfect square, which implies that $4m^4 + 4m^2 - 4m + 5$ is a perfect square.

Since

$$4m^4 + 4m^2 - 4m + 5 = (2m^2)^2 + 4m(m - 1) + 5 > (2m^2)^2,$$

$$4m^4 + 4m^2 - 4m + 5 = (2m^2 + 1)^2 - 4(m - 1) \le (2m^2 + 1)^2,$$

so it must be

$$4m^4 + 4m^2 - 4m + 5 = (2m^2 + 1)^2.$$

Hence $m = 1$.

When $m = 1$, we have $n(n + 2) = 8$, and thus $n = 2$.

Summarizing, we know that the non-negative integer solution is $(m, n) = (1, 2)$.

(4) *Method of congruence*

If an indeterminate equation $F(x_1, x_2, \ldots, x_m) = 0$ has an integer solution $(x_1, x_2, \ldots, x_m)$, then for any positive integer n,

$$F(x_1, x_2, \ldots, x_m) \equiv 0 (\mathrm{mod}\, n).$$

So we can use congruence tools to simplify indeterminate equations. Of course, in more situations, a specific n is taken to derive a contradiction, in order to prove (in some cases) that an indeterminate equation has no solution.

Example 3.7. Find positive integer solutions (x, y) to the equation $5^x - 3^y = 2$.

Solution. When $y = 1$, we have $x = 1$.

Now suppose (x, y) is a positive integer solution with $y \geq 2$.

Both sides modulo 4 leads to $1 - (-1)^y \equiv 2 (\mathrm{mod}\, 4)$, and thus y is odd.

From both sides modulo 9 we derive $5^x \equiv 2 (\mathrm{mod}\, 9)$. Note that for any non-negative integer t,

$$5^{6t} \equiv 1 (\mathrm{mod}\, 9), \quad 5^{6t+1} \equiv 5 (\mathrm{mod}\, 9), \quad 5^{6t+2} \equiv 7 (\mathrm{mod}\, 9),$$

$$5^{6t+3} \equiv 8 (\mathrm{mod}\, 9), \quad 5^{6t+4} \equiv 4 (\mathrm{mod}\, 9), \quad 5^{6t+5} \equiv 2 (\mathrm{mod}\, 9).$$

Hence we can suppose $x = 6t + 5$.

Now we consider the remainder of 5^x and 3^y divided by 7.

Note that $5^6 \equiv 1 (\mathrm{mod}\, 7)$, so $5^x \equiv (5^6)^t \cdot 5^5 \equiv 5^5 \equiv 3 (\mathrm{mod}\, 7)$. On the other hand, $3^6 \equiv 1 (\mathrm{mod}\, 7)$, so $3^y \equiv 3^1, 3^3, 3^5 (\mathrm{mod}\, 7)$, i.e., $3^y \equiv 3, 6, 5 (\mathrm{mod}\, 7)$.

Therefore, $5^x - 3^y \equiv 0, -3, -2 (\mathrm{mod}\, 7)$, which is a contradiction to $5^x - 3^y = 2$.

As a conclusion, the original equation has only one positive integer solution $(x, y) = (1, 1)$.

(5) *Prime factor exponent analysis*

Let n be a positive integer, p be a prime. Then there exists one and only one non-negative integer α such that $p^\alpha | n$ and $p^{\alpha+1} \nmid n$. This α is denoted by $v_p(n)$.

The following properties are useful:

(i) If p is a prime, then for any non-zero rational number m, n,

$$v_p(mn) = v_p(m) + v_p(n), \quad v_p\left(\frac{m}{n}\right) = v_p(m) - v_p(n).$$

(ii) If p is a prime, then for any positive integers $n_1, n_2, \ldots, n_k$,

$$v_p((n_1, n_2, \ldots, n_k)) = \min_{1 \le i \le k} v_p(n_i), \quad v_p([n_1, n_2, \ldots, n_k]) = \max_{1 \le i \le k} v_p(n_i).$$

Example 3.8. Prove that equation $a^4 + b^3 = a^3 b^2$ has no positive integer solution (a, b).

Proof. We prove by contradiction. Assume (a, b) is a positive integer solution.

Consider any prime factor p of a or b. Suppose $v_p(a) = x$, $v_p(b) = y$, where x, y are non-negative integers and at least one of them is not 0. Then

$$v_p(a^4) = 4x, \quad v_p(b^3) = 3y, \quad v_p(a^3 b^2) = 3x + 2y. \tag{1}$$

Denote $m = \min\{4x, 3y, 3x + 2y\}$.

If only one of $4x$, $3y$, $3x + 2y$ is equal to m, then the other two must be greater than m, so one of the three terms a^4, b^3, $a^3 b^2$ of the original equation is divisible by p^m and not divisible by p^{m+1}, while the other two terms are divisible by p^{m+1}, which is a contradiction.

Hence at least two of $4x$, $3y$, $3x + 2y$ are equal to m. Evidently $m > 0$ (otherwise $m = 0$ and both x, y are 0, a contradiction).

If $4x = 3x + 2y = m$, then $y = \frac{m}{8}$, so $3y = \frac{3m}{8} < m$, a contradiction to the definition of m.

If $3y = 3x + 2y = m$, then $x = \frac{m}{9}$, so $4x = \frac{4m}{9} < m$, also a contradiction to the definition of m.

Therefore, it must be $4x = 3y = m$ (since $3x + 2y = \frac{17}{12}m > m$, we have no contradiction to the definition of m here).

From the above analysis, for any prime factor p of a or b,

$$v_p(a) : v_p(b) = 3 : 4. \tag{2}$$

This implies that there exists a positive integer u such that

$$a = u^3, \quad b = u^4.$$

Substituting into the original equation, we have $u^{12} + u^{12} = u^{17}$, i.e., $u^5 = 2$, a contradiction.

As a conclusion, the original equation has no positive integer solution (a, b).

3.2 Problems and Solutions

3.2.1 *Finding solutions of indeterminate equations*

Problem 3.1 (IMO 2-1, proposed by Bulgaria). Determine all three-digit numbers N having the property that N is divisible by 11, and $\frac{N}{11}$ is equal to the sum of the squares of the digits of N.

Solution. Let the three-digit number N be $\overline{abc} = 100a + 10b + c$ with $a \neq 0$. The given conditions are equivalent to $11|a - b + c$ and $100a + 10b + c = 11(a^2 + b^2 + c^2)$. As none of a, b, c can exceed 9, we will discuss the following two cases.

(i) $a - b + c = 0$, which leads to $a^2 + (a + c)^2 + c^2 = 10a + c$ and $a + c \leq 9$. If $a + c > 5$, then $a^2 + (a + c)^2 + c^2 = 2a(a + c) + 2c^2 > 10a + c$, a contradiction. So $a + c \leq 5$.

Note that $a^2 + (a + c)^2 + c^2$ is even, so c is even, and then c can be 0, 2, 4. By checking these three numbers, we can easily find that the only solution is $c = 0$. Then $a = 5$, and the original three-digit number is 550 in this case.

(ii) $a - b + c = 11$, which leads to $a^2 + (a + c - 11)^2 + c^2 = 10a + c - 10$ and $a + c \geq 11$. If $a + c > 11$, then

$$a^2 + (a + c - 11)^2 + c^2 + 10$$

$$\geq a^2 + 1 + (12 - a)^2 + 10$$

$$= 2a^2 - 24a + 155$$

$$= 2\left(a - \frac{17}{2}\right)^2 + 10a + \frac{21}{2} > 10a + c,$$

a contradiction. So $a + c = 11$.

Now that $a^2 + (11 - a)^2 = 9a + 1$, the solution is $a = 8$. Then $c = 3$, and the original three-digit number is 803 in this case.

In summary, the only two numbers satisfying the given conditions are 550 and 803.

Remark. For number theory problems discussing decimal digits of integers, we usually write the integers into decimal representation and turn the problem into a discussion on some kind of indeterminate equation. Then in this problem, we could obtain two solutions by inequality scaling and congruence constraints.

【Score Situation】 This particular problem saw the following distribution of scores among contestants: 3 contestants scored 8 points, no contestant scored 7 points, 2 contestants scored 6 points, no contestant scored 5 points, no contestant scored 4 points, no contestant scored 3 points, 1 contestant scored 2 points, no contestant scored 1 point, and 4 contestants scored 0 point. The average score of this problem is 3.800, indicating that it was relatively straightforward.

Among the top five teams in the team scores, the Czechoslovakia team achieved a total score of 257 points, the Hungary team achieved a total score of 248 points, the Romania team achieved a total score of 248 points, the Bulgaria team achieved a total score of 175 points, and the German Democratic Republic team achieved a total score of 38 points.

The gold medal cutoff for this IMO was set at 40 points (with 4 contestants earning gold medals), the silver medal cutoff was 37 points (with 4 contestants earning silver medals), and the bronze medal cutoff was 33 points (with 4 contestants earning bronze medals).

In this IMO, no contestant achieved a perfect score of 44 points.

Problem 3.2 (IMO 4-1, proposed by Poland). Find the smallest natural number n which has the following properties:

(a) Its decimal representation has 6 as the last digit.
(b) If the last digit 6 is erased and placed in front of the remaining digits, then the resulting number is four times as large as the original number n.

Solution 1. Suppose $n = 10N + 6$, where $10^{k-1} \leq N < 10^k$ with $k \in \mathbf{N}^*$. Then by the condition,

$$4(10N + 6) = 6 \cdot 10^k + N,$$

i.e.,

$$39N = 6(10^k - 4).$$

Hence $13 | 10^k - 4$. The least positive integer satisfying $(-3)^k \equiv 4 \pmod{13}$ is $k = 5$. Then

$$N = \frac{2(10^k - 4)}{13} = 15384,$$

so the desired number is $n = 153846$.

Solution 2. Suppose $n = \overline{a_k a_{k-1} \cdots a_2 6}$. Then by the condition,

$$\overline{6 a_k a_{k-1} \cdots a_2} = 4 \cdot \overline{a_k a_{k-1} \cdots a_2 6}.$$

Since the last digit of n is 6, then $a_2 = 4$. Again by $a_2 = 4$ we get $a_3 = 8$. The digits can be obtained by a successive multiplication until 6 first occurs in a digit of $4n$. By this method, we can successively get $a_4 = 3$, $a_5 = 5$, and $a_6 = 1$. So the smallest value of n is 153846.

【Score Situation】 This particular problem saw the following distribution of scores among contestants: 14 contestants scored 6 points, 1 contestant scored 5 points, no contestant scored 4 points, no contestant scored 3 points, 1 contestant scored 2 points, 1 contestant scored 1 point, and no contestant scored 0 point. The average score of this problem is 5.412, indicating that it was simple.

Among the top five teams in the team scores, the Hungary team achieved a total score of 289 points, the Soviet Union team achieved a total score of 263 points, the Romania team achieved a total score of 257 points, the Czechoslovakia team achieved a total score of 212 points, and the Poland team achieved a total score of 212 points.

The gold medal cutoff for this IMO was set at 41 points (with 4 contestants earning gold medals), the silver medal cutoff was 34 points (with 12 contestants earning silver medals), and the bronze medal cutoff was 29 points (with 15 contestants earning bronze medals).

In this IMO, only one contestant achieved a perfect score of 46 points, namely Iosif Bernstein from the Soviet Union.

Problem 3.3 (IMO 19-5, proposed by the German Democratic Republic). Let a and b be positive integers. When $a^2 + b^2$ is divided by $a + b$, the quotient is q and the remainder is r. Find all pairs (a, b) such that $q^2 + r = 1977$.

Solution. Since r is the remainder, we know $0 \le r < a + b$. Then $q^2 \le 1977$, so $q \le 44$.

If $q \le 43$, then

$$r = 1977 - q^2 \ge 1977 - 43^2 = 128 > 2(q+1).$$

Therefore,

$$\frac{1}{2}(a+b)^2 \le a^2 + b^2$$

$$= q(a+b) + r$$

$$< q(a + b) + (a + b)$$

$$< \frac{r}{2}(a + b)$$

$$< \frac{1}{2}(a + b)^2,$$

a contradiction. Hence, it must be $q = 44$, which implies $r = 1977 - 44^2 = 41$.

Next, we substitute 44 and 41 for q and r to get the equation of a and b:

$$a^2 + b^2 = 44(a + b) + 41.$$

Equivalently,

$$(a - 22)^2 + (b - 22)^2 = 1009.$$

By checking all the square numbers less than 1009, we can find the only way to write 1009 as the sum of two square numbers is that $1009 = 15^2 + 28^2$. As a result, all pairs (a, b) satisfying the given conditions are $(a, b) = (7, 50), (37, 50), (50, 7), (50, 37)$.

Remark. As we have emphasized above, for problems that are essentially indeterminate equations, common methods include congruence analysis and inequality scaling. When we saw a "remainder", we chose to utilize its relation with the divisor to take certain scaling techniques. Then we found that q must be 44, and the rest is not difficult.

【Score Situation】 This particular problem saw the following distribution of scores among contestants: 11 contestants scored 7 points, 5 contestants scored 6 points, 1 contestant scored 5 points, 1 contestant scored 4 points, 1 contestant scored 3 points, 1 contestant scored 2 points, 7 contestants scored 1 point, and 10 contestants scored 0 point. The average score of this problem is 3.459, indicating that it was relatively straightforward.

Among the top five teams in the team scores, the United States team achieved a total score of 202 points, the Soviet Union team achieved a total score of 192 points, the Hungary team achieved a total score of 190 points, the United Kingdom team achieved a total score of 190 points, and the Netherlands team achieved a total score of 185 points.

The gold medal cutoff for this IMO was set at 34 points (with 13 contestants earning gold medals), the silver medal cutoff was 24 points (with 29 contestants earning silver medals), and the bronze medal cutoff was 17 points (with 35 contestants earning bronze medals).

In this IMO, a total of five contestants achieved a perfect score of 40 points.

Problem 3.4 (IMO 22-3, proposed by the Netherlands). Determine the maximum value of $m^2 + n^2$, where m and n are integers satisfying $m, n \in \{1, 2, \ldots, 1981\}$ and $(n^2 - mn - m^2)^2 = 1$.

Solution. If a solution satisfies $m = n$, then $m^4 = n^4 = 1$, and it can only be $m = n = 1$.

Now we consider those solutions when $m \neq n$. Since $n(n - m) = m^2 \pm 1 \geq 0$, we have $n(n - m) > 0$, so $n > m$. Observe that

$$m^2 - m(n - m) - (n - m)^2 = m^2 + mn - n^2 = \mp 1,$$
$$(n + m)^2 - n(n + m) - n^2 = m^2 + mn - n^2 = \mp 1.$$

It means that $(n - m, m)$ and $(n, n + m)$ are also solutions to the original equation.

Then for any solution where m and n are not equal, by $(m, n) \to (n - m, m)$ we can find a sequence of solutions while m, n get smaller and smaller, until the sequence terminates at $m = n$, which has been proved to be $(m, n) = (1, 1)$. Note that this construction of the sequence is reversible, in other words, we can run it backwards from $(1, 1)$ by $(m, n) \to (n, n + m)$ to get all the solutions.

Therefore, the sequence of solutions is unique, and the two numbers in every solution (m, n) are some consecutive terms of the Fibonacci sequence:

$$1, 1, 2, 3, 5, 8, 13, 21, 34, 55, 89, 144, 233, 377, 610, 987, 1597, \ldots.$$

We can see that the largest such pair not exceeding 1981 is 987 and 1597, so the maximum value is $m^2 + n^2 = 987^2 + 1597^2 = 3524578$.

Remark. There are two points worth our attention. One is that the sequence of solutions in fact comes from Vieta's formula if we treat $n^2 - mn - m^2 = \pm 1$ as a quadratic equation of m and n respectively. This method of using Vieta's formula is quite common when dealing with quadratic indeterminate equations. Another point is that two consecutive terms of the Fibonacci sequence $\{F_n\}_{n \geq 0}$ satisfy

$$F_{n+1}^2 - F_n F_{n+1} - F_n^2 = \pm 1,$$

which are exactly all the solutions to the original problem.

【Score Situation】 This particular problem saw the following distribution of scores among contestants: 32 contestants scored 7 points, 2 contestants scored 6 points, 3 contestants scored 5 points, no contestant scored 4 points, 2 contestants scored 3 points, 3 contestants scored 2 points, 3 contestants scored 1 point, and 6 contestants scored 0 point. The average score of this problem is 5.216, indicating that it was simple.

Among the top five teams in the team scores, the scores of this problem are as follows: The United States team scored 45 points (with a total team score of 314 points), the Germany team scored 43 points (with a total team score of 312 points), the United Kingdom team scored 45 points (with a total team score of 301 points), the Austria team scored 36 points (with a total team score of 290 points), and the Bulgaria team scored 36 points (with a total team score of 287 points).

The gold medal cutoff for this IMO was set at 41 points (with 36 contestants earning gold medals), the silver medal cutoff was 34 points (with 37 contestants earning silver medals), and the bronze medal cutoff was 26 points (with 30 contestants earning bronze medals).

In this IMO, a total of 26 contestants achieved a perfect score of 42 points.

Problem 3.5 (IMO 38-5, proposed by Czech Republic). Find all pairs (a, b) of integers $a, b \geq 1$ that satisfy the equation $a^{b^2} = b^a$.

Solution. For a solution, if either a or b is equal to 1, then obviously $(a, b) = (1, 1)$.

Now suppose $a, b \geq 2$. Let $t = \frac{b^2}{a} > 0$. Substituting into the given condition, we get $a^{at} = b^a$, so $b = a^t$, and

$$at = b^2 = a^{2t}, \quad t = a^{2t-1}.$$

Assume $t \geq 1$. Then $2t - 1 \geq 1$, and thus

$$t = a^{2t-1} \geq (1+1)^{2t-1} \geq 1 + (2t-1) > t,$$

a contradiction. Hence $0 < t < 1$.

Let a rational number $k = \frac{1}{t} = \frac{a}{b^2} > 1$. Then for $b = a^t$ we know $a = b^k$. Since $b^{kb^2} = b^{b^k}$,

$$k = b^{k-2}. \tag{$*$}$$

Note that $b^{k-2} = k > 1$, so $k > 2$. Suppose $k = \frac{p}{q}$, where p, q are coprime positive integers. Then $p > 2q$. By $(*)$,

$$\left(\frac{p}{q}\right)^q = k^q = b^{p-2q}$$

is an integer, which implies $q|p$, and together with $(p,q) = 1$ we know $q = 1$, i.e., k is an integer greater than 2.

   If $k = 3$, then $b = 3$, and $(a, b) = (27, 3)$, a solution.
   If $k = 4$, then $b = 2$, and $(a, b) = (16, 2)$, also a solution.
   However, if $k \geq 5$, then by $(*)$,

$$k \geq 2^{k-2} = (1+1)^{k-2} \geq 1 + k - 2 + \frac{(k-2)(k-3)}{2} > k,$$

which is a contradiction.

   As a conclusion, all the desired pairs are $(a, b) = (1,1), (16,2), (27,3)$.

Remark. Again we used the scaling technique in this solution, since we usually can obtain valuable information from scaling for non-homogeneous indeterminate equations. In fact, this problem also has other solutions which focus more on congruence analysis. Specifically, we can check the greatest common divisor of a and b to obtain equation $(*)$.

【Score Situation】 This particular problem saw the following distribution of scores among contestants: 152 contestants scored 7 points, 14 contestants scored 6 points, 19 contestants scored 5 points, 21 contestants scored 4 points, 31 contestants scored 3 points, 36 contestants scored 2 points, 51 contestants scored 1 point, and 136 contestants scored 0 point. The average score of this problem is 3.354, indicating that it was relatively straightforward.

   Among the top five teams in the team scores, the scores of this problem are as follows: The China team scored 41 points (with a total team score of 223 points), the Hungary team scored 42 points (with a total team score of 219 points), the Iran team scored 42 points (with a total team score of 217 points), the Russia team scored 42 points (with a total team score of 202 points), and the United States team scored 38 points (with a total team score of 202 points).

   The gold medal cutoff for this IMO was set at 35 points (with 39 contestants earning gold medals), the silver medal cutoff was 25 points (with 70 contestants earning silver medals), and the bronze medal cutoff was 15 points (with 122 contestants earning bronze medals).

   In this IMO, a total of four contestants achieved a perfect score of 42 points.

Problem 3.6 (IMO 44-2, proposed by Bulgaria). Find all pairs (a, b) of positive integers such that $\frac{a^2}{2ab^2 - b^3 + 1}$ is a positive integer.

Solution 1. Let (a, b) be a pair of positive integers satisfying the condition. Since $k = \frac{a^2}{2ab^2 - b^3 + 1} > 0$, we have $2ab^2 - b^3 + 1 > 0$ or $a > \frac{b}{2} - \frac{1}{2b^2}$, and hence $a \geq \frac{b}{2}$. Using this, we infer from $k \geq 1$, or $a^2 \geq b^2(2a - b) + 1$, that

$a^2 > b^2(2a - b) \geq 0.$

$$\text{Hence } a > b \text{ or } 2a = b. \tag{1}$$

Now consider the two solutions a_1, a_2 of the equation

$$a^2 - 2kb^2a + k(b^3 - 1) = 0 \tag{2}$$

for any fixed positive integers k and b, and assume that one of them is an integer. Then the other is also an integer because $a_1 + a_2 = 2kb^2$. We may assume that $a_1 \geq a_2$, and we have $a_1 \geq kb^2 > 0$. Furthermore, since $a_1 a_2 = k(b^3 - 1)$,

$$0 \leq a_2 = \frac{k(b^3 - 1)}{a_1} \leq \frac{k(b^3 - 1)}{kb^2} < b.$$

Together with (1), we conclude that $a_2 = 0$ or $a_2 = \frac{b}{2}$ (in the latter case b must be even).

If $a_2 = 0$, then $b^3 - 1 = 0$, and hence $a_1 = 2k$ and $b = 1$.

If $a_2 = \frac{b}{2}$, then $k = \frac{b^2}{4}$ and $a_1 = \frac{b^4}{2} - \frac{b}{2}$.

Therefore, the only solutions are

$$(a, b) = (2l, 1), (l, 2l), (8l^4 - l, 2l)$$

for some positive integer l. All of these pairs satisfy the given condition.

Solution 2. If $b = 1$, then it follows from the given condition that a must be even.

Let $\frac{a^2}{2ab^2 - b^3 + 1} = k$. If $b > 1$, then there are two solutions to the equation (2) and one of them is a positive integer. Thus the discriminant Δ of the equation (2) is a perfect square, that is, $\Delta = 4k^2b^4 - 4k(b^3 - 1)$ is a perfect square.

Note that, if $b \geq 2$, then

$$(2kb^2 - b - 1)^2 < \Delta < (2kb^2 - b + 1)^2. \tag{3}$$

The proof is given as follows:

$$\Delta - (2kb^2 - b - 1)^2 = 4kb^2 - b^2 - 2b + 4k - 1$$
$$= (4k - 1)(b^2 + 1) - 2b$$
$$\geq 2(4k - 1)b - 2b > 0,$$

$$(2kb^2 - b + 1)^2 - \Delta = 4kb^2 - 4k + (b-1)^2$$
$$= 4k(b^2 - 1) - (b-1)^2$$
$$> (4k-1)(b^2 - 1) > 0.$$

This completes the proof of (3).

Since Δ is a perfect square, it follows from (3) that

$$\Delta = 4k^2b^4 - 4k(b^3 - 1) = (2kb^2 - b)^2.$$

Then $b^2 = 4k$, and hence b must be even. Let $b = 2l$. We have $k = l^2$. Together with (2), we obtain $a = l$ or $8l^4 - l$.

Therefore, the only solutions are

$$(a, b) = (2l, 1), (l, 2l), (8l^4 - l, 2l)$$

for some positive integer l. All of these pairs satisfy the given condition.

Remark. Note that we can obtain a quadratic equation of a after simple transformation. Then it is common to use the discriminant or Vieta's formula. Fortunately, both methods can reach the final solution.

【Score Situation】 This particular problem saw the following distribution of scores among contestants: 67 contestants scored 7 points, 1 contestant scored 6 points, 20 contestants scored 5 points, no contestant scored 4 points, 93 contestants scored 3 points, 40 contestants scored 2 points, 119 contestants scored 1 point, and 117 contestants scored 0 point. The average score of this problem is 2.304, indicating that it had a certain level of difficulty.

Among the top five teams in the team scores, the scores of this problem are as follows: The Bulgaria team scored 34 points (with a total team score of 227 points), the China team scored 40 points (with a total team score of 211 points), the United States team scored 36 points (with a total team score of 188 points), the Vietnam team scored 38 points (with a total team score of 172 points), and the Russia team scored 32 points (with a total team score of 167 points).

The gold medal cutoff for this IMO was set at 29 points (with 37 contestants earning gold medals), the silver medal cutoff was 19 points (with 69 contestants earning silver medals), and the bronze medal cutoff was 13 points (with 104 contestants earning bronze medals).

In this IMO, only three contestants achieved a perfect score of 42 points, namely Bảo Lê Hùng Việt and Trọng Cảnh Nguyễn from Vietnam, and Yunhao Fu from China.

Problem 3.7 (IMO 47-4, proposed by the United States). Determine all pairs (x, y) of integers such that

$$1 + 2^x + 2^{2x+1} = y^2.$$

Solution. Evidently, for any solution (x, y), we have $x \geq 0$, and $(x, -y)$ is also a solution. When $x = 0$, obviously the only two solutions are $(0, 2)$ and $(0, -2)$.

Next we consider the solutions (x, y) where $x, y > 0$. Then the original equation is equivalent to

$$2^x(1 + 2^{x+1}) = (y - 1)(y + 1).$$

Hence, $y - 1$ and $y + 1$ are both even; moreover, only one of them is divisible by 4. Consequently, $x \geq 3$, and one of $y - 1$ and $y + 1$ is divisible by 2^{x-1} but not 2^x.

If $y = 2^{x-1}m + 1$, where m is odd, then

$$1 + 2^{x+1} = 2^{x-2}m^2 + m,$$

so that

$$1 - m = 2^{x-2}(m^2 - 8).$$

Therefore, $m^2 < 8$, which leads to $m = 1$, implying that the equality does not hold.

Otherwise, $y = 2^{x-1}m - 1$, where m is odd, and then

$$1 + 2^{x+1} = 2^{x-2}m^2 - m,$$

so that

$$1 + m = 2^{x-2}(m^2 - 8) \geq 2(m^2 - 8),$$

i.e.,

$$2m^2 - m - 17 \leq 0.$$

Together with $m^2 > 8$ we know that the only odd number satisfying these two inequalities is $m = 3$, and the corresponding solution is $x = 4$ and $y = 23$.

Thus we have the complete list of solutions $(x, y) = (0, 2), (0, -2),$ $(4, 23), (4, -23)$.

Remark. For such an indeterminate equation, it is common to discuss the divisibility of factors after using the formula for the difference of squares.

This method is also widely used in the analysis of Pythagorean triples. Then using the obtained conclusions, we can write the forms of y and get the final solution after some discussion.

【Score Situation】 This particular problem saw the following distribution of scores among contestants: 248 contestants scored 7 points, 68 contestants scored 6 points, 7 contestants scored 5 points, 6 contestant scored 4 points, 38 contestant scored 3 points, 59 contestants scored 2 points, 54 contestants scored 1 point, and 18 contestants scored 0 point. The average score of this problem is 4.998, indicating that it was simple.

Among the top five teams in the team scores, the scores of this problem are as follows: The China team scored 41 points (with a total team score of 214 points), the Russia team scored 42 points (with a total team score of 174 points), the South Korea team scored 42 points (with a total team score of 170 points), the Germany team scored 42 points (with a total team score of 157 points), and the United States team scored 41 points (with a total team score of 154 points).

The gold medal cutoff for this IMO was set at 28 points (with 42 contestants earning gold medals), the silver medal cutoff was 19 points (with 89 contestants earning silver medals), and the bronze medal cutoff was 15 points (with 122 contestants earning bronze medals).

In this IMO, only three contestants achieved a perfect score of 42 points, namely Zhiyu Liu from China, Iurie Boreico from Moldova and Alexander Magazinov from Russia.

Problem 3.8 (IMO 56-2, proposed by Serbia). Determine all triples (a, b, c) of positive integers such that each of the numbers

$$ab - c, \quad bc - a, \quad ca - b$$

is a power of 2 (a power of 2 is an integer of the form 2^n, where n is a non-negative integer).

Solution. Let (a, b, c) be such a triple. Then $a = 1$ would imply that both $b - c$ and $c - b$ are powers of 2, which is impossible. So $a \geq 2$; similarly, $b \geq 2$ and $c \geq 2$. We discuss two cases.

Case 1. At least two of a, b, and c are the same.
Without loss of generality, let $a = b$. Then $ac - b = a(c - 1)$ is a power of 2, so are both a and $c - 1$. Let $a = 2^s$ and $c = 1 + 2^t$, where $s \geq 1$, $t \geq 0$. Then $ab - c = 2^{2s} - 2^t - 1$ is a power of 2. When $t > 0$, the integer $2^{2s} - 2^t - 1$ is odd , so $2^{2s} - 2^t - 1 = 1$. Then $2^t \equiv 2 \pmod 4$, so $t = 1$ and

$s = 1$, $a = b = 2$, and $c = 3$. When $t = 0$, the integer $2^{2s} - 2$ is a power of 2, so we must have $s = 1$ and $a = b = c = 2$. It is easy to verify that both triples $(2, 2, 2)$ and $(2, 2, 3)$ satisfy our requirement.

Case 2. a, b, and c are pairwise distinct.
Without loss of generality, suppose $2 \leq a < b < c$. By our assumption, there are non-negative integers α, β, and γ such that

$$bc - a = 2^{\alpha}, \tag{1}$$

$$ca - b = 2^{\beta}, \tag{2}$$

$$ab - c = 2^{\gamma}. \tag{3}$$

It is easy to see that $\alpha > \beta > \gamma \geq 0$. We discuss two sub-cases based on the value of a.

Case 2.1. $a = 2$.
We first prove that $\gamma = 0$. Otherwise, by (3), c is even; and by (2), b is even; thus, the left hand side of (1), $bc - a \equiv 2 \pmod 4$; yet $2^{\alpha} \equiv 0 \pmod 4$, a contradiction.

So $\gamma = 0$, and (3) becomes $c = 2b - 1$. Combining with (2), we get $3b - 2 = 2^{\beta}$. Take modulo 3 we see that β is even. If $\beta = 2$, then $b = 2 = a$, which contradicts our assumption in this case. If $\beta = 4$, we get $b = 6$ and $c = 11$; and it is easy to verify that $(2, 6, 11)$ is a triple that satisfies our requirement. If $\beta \geq 6$, then $b = \frac{1}{3}(2^{\beta} + 2)$.

Substituting into (1) gives

$$9 \times 2^{\alpha} = 9(bc - a)$$
$$= 9b(2b - 1) - 18$$
$$= (3b - 2)(6b + 1) - 16$$
$$= 2^{\beta}(2^{\beta+1} + 5) - 16.$$

Since $\alpha > \beta \geq 6$, we have $2^7 | 9 \cdot 2\alpha$; however the right hand side of the above only divides 2^4 but not 2^5, a contradiction.

Case 2.2. $a \geq 3$.
Adding (1) and (2), we get

$$(a + b)(c - 1) = 2^{\alpha} + 2^{\beta}.$$

Subtracting (2) from (1), we get

$$(b - a)(c + 1) = 2^{\alpha} - 2^{\beta}.$$

One of $c+1$ and $c-1$ is not a multiple of 4, so $2^{\beta-1}|a+b$ or $2^{\beta-1}|b-a$. From

$$2^\beta = ac - b \geq 3c - b > 2c$$

we get $b < c < 2^{\beta-1}$, so $0 < b-a < 2^{\beta-1}$, and it is impossible for $2^{\beta-1}|b-a$. Therefore $2^{\beta-1}|a+b$, and since $a+b < 2b < 2^\beta$, we must have $a+b = 2^{\beta-1}$.

Substituting into (2) gives

$$ac - b = 2^\beta = 2(a+b),$$

i.e., $a(c-2) = 3b$. When $a \geq 4$, we have $b \geq 5$, and

$$a(c-2) \geq 4(c-2) \geq 4(b-1) > 3b,$$

a contradiction. So $a = 3$ and $c - 2 = b$. Therefore $b = 2^{\beta-1} - 3$ and $c = 2^{\beta-1} - 1$. Substituting into (3) leads to

$$2^\gamma = ab - c = 3(2^{\beta-1} - 3) - (2^{\beta-1} - 1) = 2^\beta - 8,$$

which implies $\beta = 4$, $b = 5$, and $c = 7$. It is easy to verify that $(3, 5, 7)$ satisfies our requirement.

So there are 16 such triples — $(2, 2, 2)$, the 3 permutations of $(2, 2, 3)$, the 6 permutations of $(2, 6, 11)$, and the 6 permutations of $(3, 5, 7)$.

【Score Situation】 This particular problem saw the following distribution of scores among contestants: 31 contestants scored 7 points, 14 contestants scored 6 points, 13 contestants scored 5 points, 8 contestants scored 4 points, 27 contestants scored 3 points, 77 contestants scored 2 points, 151 contestants scored 1 point, and 256 contestants scored 0 point. The average score of this problem is 1.359, indicating that it was relatively challenging.

Among the top five teams in the team scores, the scores of this problem are as follows: The United States team scored 32 points (with a total team score of 185 points), the China team scored 36 points (with a total team score of 181 points), the South Korea team scored 15 points (with a total team score of 161 points), the North Korea team scored 25 points (with a total team score of 156 points), and the Vietnam team scored 21 points (with a total team score of 151 points).

The gold medal cutoff for this IMO was set at 26 points (with 39 contestants earning gold medals), the silver medal cutoff was 19 points (with 100 contestants earning silver medals), and the bronze medal cutoff was 14 points (with 143 contestants earning bronze medals).

In this IMO, only one contestant achieved a perfect score of 42 points, namely Zhuo Qun Alex Song from Canada.

Problem 3.9 (IMO 60-4, proposed by El Salvador). Find all pairs (k, n) of positive integers such that

$$k! = (2^n - 1)(2^n - 2)(2^n - 4) \cdots (2^n - 2^{n-1}).$$

Solution. Denote the expression on the right side as R_n. Then

$$v_2(R_n) = \sum_{i=0}^{n-1} v_2(2^n - 2^i) = \sum_{i=0}^{n-1} i = \frac{1}{2}n(n-1).$$

It is well known that $v_2(k!) = k - S_2(k)$, where $S_2(k)$ denotes the sum of digits in the binary representation of k. Hence

$$k - S_2(k) = \frac{1}{2}n(n-1),$$

and $k = \frac{1}{2}n(n-1) + S_2(k) \geq \frac{1}{2}n(n-1) + 1$.

Checking $n = 1, 2, 3, 4$ directly, we get two solutions $(k, n) = (1, 1), (3, 2)$. For $n \geq 5$, we claim that

$$\left(\frac{1}{2}n(n-1) + 1\right)! > 2^{n^2},$$

while $R_n \leq (2^n)^n = 2^{n^2}$, so the equation has no solution.

For $n = 5$, we directly verify that $11! = 39916800 > 2^{25} = 33554432$.
For $n > 5$,

$$\left(\frac{1}{2}n(n-1) + 1\right)! = 11! \cdot \prod_{i=12}^{\frac{1}{2}n(n-1)+1} i > 2^{25} \cdot 8^{\frac{1}{2}n(n-1)-10}$$

$$= 2^{25 + \frac{3}{2}n(n-1) - 30} \geq 2^{\frac{3}{2}n(n-1) - 5} \geq 2^{n^2}.$$

In summary, $(k, n) = (1, 1), (3, 2)$ are the only pairs that satisfy the equation.

Remark. The difference in form between the two sides of the given equation is significant, so it is natural to consider inequality scaling in order to estimate the range of k and n. The most essential constraint of the size relation between k and n is that the exponent of 2 on the right hand side is significantly large. By analyzing the power of 2, we can give a lower bound of k relative to n. Then the last step is to prove that this k will make the overall left hand side larger than the right hand side.

【Score Situation】This particular problem saw the following distribution of scores among contestants: 257 contestants scored 7 points, 47 contestants scored 6 points, 19 contestants scored 5 points, 13 contestants scored 4 points, 7 contestants scored 3 points, 4 contestants scored 2 points, 63 contestants scored 1 point, and 211 contestants scored 0 point. The average score of this problem is 3.736, indicating that it was relatively straightforward.

Among the top five teams in the team scores, the scores of this problem are as follows: The United States team scored 42 points (with a total team score of 227 points), the China team scored 41 points (with a total team score of 227 points), the South Korea team scored 42 points (with a total team score of 226 points), the North Korea team scored 42 points (with a total team score of 187 points), and the Thailand team scored 42 points (with a total team score of 185 points).

The gold medal cutoff for this IMO was set at 31 points (with 52 contestants earning gold medals), the silver medal cutoff was 24 points (with 94 contestants earning silver medals), and the bronze medal cutoff was 17 points (with 156 contestants earning bronze medals).

In this IMO, a total of 6 contestants achieved a perfect score of 42 points.

Problem 3.10 (IMO 63-5, proposed by Belgium). Find all triples (a, b, p) of positive integers with p prime and

$$a^p = b! + p.$$

Solution. There are only two such triples: $(2, 2, 2)$ and $(3, 4, 3)$. It is straightforward to check them. We shall prove that no other triples exist.

Clearly, $a > 1$. Consider three situations as follows.

(i) $a < p$. If $a \le b$, then $a|(a^p - b!)$, contradicting the assumption $1 < a < p$; if $a > b$, then $b! \le a! < a^p - p$, the second inequality only requiring $p > a > 1$.

(ii) $a > p$. Then $b! = a^p - p > p^p - p \ge p!$, $b > p$, and $a^p = b! + p$ is a multiple of p. As $b! = a^p - p$, we have $p\|b$, and $b < 2p$. If $a < p^2$, then $\frac{a}{p}$ divides $a^p, b!$, and hence divides p as well, contradicting $1 < \frac{a}{p} < p$; if $a \ge p^2$, then it is contradicting $a^p \ge (p^2)^p > (2p-1)! + p \ge b! + p$.

(iii) $a = p$. Then $b! = p^p - p$. Try $p = 2, 3, 5$ to get the two triples $(2, 2, 2)$ and $(3, 4, 3)$. Now assume $p \ge 7$. From $b! = p^p - p > p!$, it follows $b \ge p + 1$, and further

$$v_2((p+1)!) \le v_2(b!) = v_2(p^{p-1} - 1) = 2v_2(p-1) + v_2(p+1) - 1$$

$$= v_2\left(\frac{p-1}{2} \cdot (p-1) \cdot (p+1)\right).$$

Since $\frac{p-1}{2}, (p-1), (p+1)$ are distinct factors of $(p+1)!$ and $p+1 \geq 8$, there are four or more even numbers among $1, 2, \ldots, p+1$, which is impossible.

【Score Situation】This particular problem saw the following distribution of scores among contestants: 171 contestants scored 7 points, 49 contestants scored 6 points, 35 contestants scored 5 points, 15 contestants scored 4 points, 48 contestants scored 3 points, 44 contestants scored 2 points, 115 contestants scored 1 point, and 112 contestants scored 0 point. The average score of this problem is 3.520, indicating that it was relatively straightforward.

Among the top five teams in the team scores, the scores of this problem are as follows: The China team scored 42 points (with a total team score of 252 points), the South Korea team scored 41 points (with a total team score of 208 points), the United States team scored 39 points (with a total team score of 207 points), the Vietnam team scored 38 points (with a total team score of 196 points), and the Romania team scored 42 points (with a total team score of 194 points).

The gold medal cutoff for this IMO was set at 34 points (with 44 contestants earning gold medals), the silver medal cutoff was 29 points (with 101 contestants earning silver medals), and the bronze medal cutoff was 23 points (with 140 contestants earning bronze medals).

In this IMO, a total of 10 contestants achieved a perfect score of 42 points.

3.2.2 Proving properties satisfied by indeterminate equations

Problem 3.11 (IMO 23-4, proposed by the United Kingdom). Prove that if n is a positive integer such that the equation

$$x^3 - 3xy^2 + y^3 = n$$

has a solution in integers (x, y), then it has at least three such solutions.

Show that the equation has no solutions in integers when $n = 2891$.

Proof. (a) If the equation has a solution in integers (x, y), then by

$$(y-x)^3 = y^3 - 3xy^2 + 3x^2y - x^3 = y^3 - 3xy^2 + x^3 + 3x^2(y-x) + x^3,$$

we get

$$(y-x)^3 - 3(y-x)x^2 + (-x)^3 = y^3 - 3xy^2 + x^3 = n.$$

This implies that except (x, y), the pair $(y-x, -x)$ is a solution of the original equation; moreover, $(-x - (y-x), -(y-x)) = (-y, x-y)$ is also a solution.

If any two of these solutions are the same, then $x = y = 0$ and $n = 0$, a contradiction. Therefore these three solutions are different from each other, and the proposition is proved.

(b) When $n = 2891$, if equation $x^3 - 3xy^2 + y^3 = 2891$ has an integer solution (x, y), then

$$x^3 - 3xy^2 + y^3 \equiv 2 \pmod 9,$$

so x and y cannot be simultaneously divisible by 3.

If one of x, y is divisible by 3, then consider the other two solutions $(y - x, -x)$ and $(-y, x - y)$ in (a). At least one of these solutions has a pair of numbers both not divisible by 3. Then take this solution, and without loss of generality, assume that neither x nor y is divisible by 3. For this (x, y), we have $x^3 \equiv \pm 1 \pmod 9$, $y^3 \equiv \pm 1 \pmod 9$, and $-3xy^2 \equiv \pm 3 \pmod 9$. Hence

$$x^3 - 3xy^2 + y^3 \equiv 1, 3, 4, 5, 6, 8 \pmod 9,$$

which is a contradiction. So the equation has no solutions in integers when $n = 2891$.

Remark. Based on the form of the original equation, it is not difficult to think of the expansion of $(x + y)^3$, and then construct the other two solutions in (a). For (b), we discussed congruence properties. In fact, in addition to congruence modulo 9, contradictions can also be derived through congruence modulo 7 (note that 2891 can be divided by 7 but not by 7^3, so x and y cannot be divided by 7).

【Score Situation】 This particular problem saw the following distribution of scores among contestants: 44 contestants scored 7 points, 5 contestants scored 6 points, 6 contestants scored 5 points, 8 contestants scored 4 points, 14 contestants scored 3 points, 2 contestants scored 2 points, 4 contestants scored 1 point, and 36 contestants scored 0 point. The average score of this problem is 3.782, indicating that it was relatively straightforward.

Among the top five teams in the team scores, the scores of this problem are as follows: The Germany team scored 21 points (with a total team score of 145 points), the Soviet Union team scored 21 points (with a total team score of 137 points), the United States team scored 24 points (with a total team score of 136 points), the German Democratic Republic team scored 21 points (with a total team score of 136 points), and the Vietnam team scored 24 points (with a total team score of 133 points).

The gold medal cutoff for this IMO was set at 37 points (with 10 contestants earning gold medals), the silver medal cutoff was 30 points (with 20 contestants earning silver medals), and the bronze medal cutoff was 21 points (with 31 contestants earning bronze medals).

In this IMO, only three contestants achieved a perfect score of 42 points, namely Bruno Haible from Germany, Grigori Perelman from the Soviet Union and Lê Tự Quốc Thắng from Vietnam.

Problem 3.12 (IMO 27-1, proposed by Germany). Let d be any positive integer not equal to 2, 5, or 13. Show that one can find distinct a and b in the set $\{2, 5, 13, d\}$ such that $ab - 1$ is not a perfect square.

Proof. We need only to show that at least one of $2d-1$, $5d-1$, and $13d-1$ is not a perfect square. Assume for contrary that

$$2d - 1 = x^2, \tag{1}$$

$$5d - 1 = y^2, \tag{2}$$

$$13d - 1 = z^2, \tag{3}$$

where x, y, and z are positive integers.

By (1), x must be odd. Then $x^2 \equiv 1 \pmod 4$, so $2d \equiv 2 \pmod 4$, and thus d is odd.

For odd d, by (2) and (3) we know y and z are even. Let $y = 2u$ and $z = 2v$. Then

$$8d = z^2 - y^2 = 4(v^2 - u^2),$$

i.e., $2d = v^2 - u^2$. Hence u and v are of the same parity, which leads to $u^2 \equiv v^2 \pmod 4$, and consequently $2d \equiv 0 \pmod 4$, so d is even. This gives the desired contradiction.

Remark. For indeterminate equation problems with specific numbers, we usually start by a simple congruence analysis on some small prime numbers. Fortunately, by directly discussing properties modulo 2, we can deduce the contradiction in this proof. Of course, modulo 2 here also includes modulo powers of 2, such as the use of the classic conclusions: even perfect square numbers are congruent to 0 modulo 4 and odd perfect square numbers are congruent to 1 modulo 4 (and modulo 8).

【Score Situation】 This particular problem saw the following distribution of scores among contestants: 68 contestants scored 7 points, 4 contestants scored 6 points, 11 contestants scored 5 points, 10 contestants scored 4 points, 33 contestants scored 3 points, 43 contestants scored 2 points, 25 contestants scored 1 point, and 16 contestants scored 0 point. The average score of this problem is 3.833, indicating that it was relatively straightforward.

Among the top five teams in the team scores, the scores of this problem are as follows: The United States team scored 33 points (with a total team score of 203 points), the Soviet Union team scored 24 points (with a total team score of 203 points), the Germany team scored 32 points (with a total team score of 196 points), the China team scored 30 points (with a total team score of 177 points), and the German Democratic Republic team scored 28 points (with a total team score of 172 points).

The gold medal cutoff for this IMO was set at 34 points (with 18 contestants earning gold medals), the silver medal cutoff was 26 points (with 41 contestants earning silver medals), and the bronze medal cutoff was 17 points (with 48 contestants earning bronze medals).

In this IMO, only three contestants achieved a perfect score of 42 points, namely Vladimir Roganov and Stanislav Smirnov from the Soviet Union, and Géza Kós from Hungary.

Problem 3.13 (IMO 29-6, proposed by Germany). Let a and b be positive integers such that $ab + 1$ divides $a^2 + b^2$. Show that $\frac{a^2+b^2}{ab+1}$ is the square of an integer.

Proof. Suppose $\frac{a^2+b^2}{ab+1} = k$ with k a positive integer. Then we shall show that k is a perfect square.

Assume, by the way of contradiction, that k is not a square of an integer. Consider integer solutions (x, y) of the equation

$$x^2 - kxy + y^2 - k = 0.$$

It has an integer solution $(x, y) = (a, b)$.

If any one of x and y is 0, then $k = x^2$ or $k = y^2$ is a perfect square, a contradiction. Hence we can assume $xy \neq 0$. If $xy \leq -1$, then $x^2 + y^2 = k(xy + 1) \leq 0$, a contradiction. So $xy > 0$.

There exists a solution satisfying $x > 0, y > 0$, like $(x, y) = (a, b)$. Among these solutions we take the one whose $x + y$ is the least, and denote it by (x, y).

Without loss of generality, assume $x \geq y$. Now we consider the original equation as a quadratic equation of x. One root of this quadratic equation is x, and we denote the other root by x'. By Vieta's formulas,

$$x + x' = ky, \tag{1}$$

$$xx' = y^2 - k. \tag{2}$$

From (1) we know that x' is an integer. Thus (x', y) is also a solution of the original equation. As $y > 0$, it holds $x' > 0$. From (2),

$$x' = \frac{y^2 - k}{x} \leq \frac{x^2 - k}{x} < x,$$

so $x' + y < x + y$, which is a contradiction to that $x + y$ is the least. As a conclusion, $\frac{a^2 + b^2}{ab + 1}$ is a perfect square.

Remark. In the previous problems, we have mentioned Vieta's formula for several times. And this proof is another standard application of Vieta's formula for indeterminate equations. The process that first pick a least solution and then use Vieta's formula to find a smaller solution for contradiction can be used in many indeterminate equations with quadratic terms and multiple variables.

【Score Situation】This particular problem saw the following distribution of scores among contestants: 11 contestants scored 7 points, 1 contestant scored 6 points, 1 contestant scored 5 points, 1 contestant scored 4 points, 5 contestants scored 3 points, 3 contestants scored 2 points, 57 contestants scored 1 point, and 189 contestants scored 0 point. The average score of this problem is 0.634, indicating that it was extremely difficult.

Among the top five teams in the team scores, the scores of this problem are as follows: The Soviet Union team scored 17 points (with a total team score of 217 points), the Romania team scored 18 points (with a total team score of 201 points), the China team scored 17 points (with a total team score of 201 points), the Germany team scored 2 points (with a total team score of 174 points), and the Vietnam team scored 7 points (with a total team score of 166 points).

The gold medal cutoff for this IMO was set at 32 points (with 17 contestants earning gold medals), the silver medal cutoff was 23 points (with 48 contestants earning silver medals), and the bronze medal cutoff was 14 points (with 65 contestants earning bronze medals).

In this IMO, a total of five contestants achieved a perfect score of 42 points.

Problem 3.14 (IMO 54-1, proposed by Japan). Prove that for any pair of positive integers k and n, there exist k positive integers $m_1, m_2, \ldots, m_k$ (not necessarily different) such that

$$1 + \frac{2^k - 1}{n} = \left(1 + \frac{1}{m_1}\right)\left(1 + \frac{1}{m_2}\right) \cdots \left(1 + \frac{1}{m_k}\right). \tag{1}$$

Proof 1. By induction on k, the case of $k = 1$ is trivial. Suppose that the proposition is true for $k = j - 1$. We show that the case of $k = j$ is also true.

Suppose n is odd, that is, $n = 2t - 1$ for some positive integer t. Note that

$$1 + \frac{2^j - 1}{2t - 1} = \frac{2(t + 2^{j-1} - 1)}{2t} \cdot \frac{2t}{2t - 1} = \left(1 + \frac{2^{j-1} - 1}{t}\right)\left(1 + \frac{1}{2t - 1}\right),$$

and by the hypothesis of induction, we can find $m_1, \ldots, m_{j-1}$ such that

$$1 + \frac{2^{j-1} - 1}{t} = \left(1 + \frac{1}{m_1}\right)\left(1 + \frac{1}{m_2}\right) \cdots \left(1 + \frac{1}{m_{j-1}}\right).$$

Thus, we need only to take $m_j = 2t - 1$.

Suppose n is even, that is, $n = 2t$ for some positive integer t. Note that

$$1 + \frac{2^j - 1}{2t} = \frac{2t + 2^j - 1}{2t + 2^j - 2} \cdot \frac{2t + 2^j - 2}{2t}$$

$$= \left(1 + \frac{1}{2t + 2^j - 2}\right)\left(1 + \frac{2^{j-1} - 1}{t}\right)$$

and $2t + 2^j - 2 > 0$. By the hypothesis of induction, we can find $m_1, \ldots, m_{j-1}$ such that

$$1 + \frac{2^{j-1} - 1}{t} = \left(1 + \frac{1}{m_1}\right)\left(1 + \frac{1}{m_2}\right) \cdots \left(1 + \frac{1}{m_{j-1}}\right).$$

Thus, we need only to take $m_{j-1} = 2t + 2^j - 2$.

Proof 2. Consider the binary expansion of the remainders of $n - 1$ and $-n \bmod 2^k$:

$$n - 1 \equiv 2^{a_1} + 2^{a_2} + \cdots + 2^{a_r} \pmod{2^k},$$

where $0 \le a_1 < a_2 < \cdots < a_r \le k - 1$, and

$$-n \equiv 2^{b_1} + 2^{b_2} + \cdots + 2^{b_s} \pmod{2^k},$$

where $0 \le b_1 < b_2 < \cdots < b_s \le k - 1$. Since $-1 \equiv 2^0 + 2^1 + \cdots + 2^{k-1} \pmod{2^k}$, we have

$$\{a_1, a_2, \ldots, a_r\} \cup \{b_1, b_2, \ldots, b_s\} = \{0, 1, \ldots, k - 1\}, \text{ and } r + s = k.$$

For $1 \le p \le r$, $1 \le q \le s$, we denote

$$S_p = 2^{a_p} + 2^{a_{p+1}} + \cdots + 2^{a_r}, \quad T_q = 2^{b_1} + 2^{b_2} + \cdots + 2^{b_q}.$$

And define $S_{r+1} = T_0 = 0$. Note that $S_1 + T_s = 2^k - 1$ and $n + T_s \equiv 0 \pmod{2^k}$. We have

$$1 + \frac{2^k - 1}{n} = \frac{n + S_1 + T_s}{n} = \frac{n + S_1 + T_s}{n + T_s} \cdot \frac{n + T_s}{n}$$

$$= \prod_{p=1}^{r} \frac{n + S_p + T_s}{n + S_{p+1} + T_s} \cdot \prod_{q=1}^{s} \frac{n + T_q}{n + T_{q-1}}$$

$$= \prod_{p=1}^{r} \left(1 + \frac{2^{a_p}}{n + S_{p+1} + T_s}\right) \cdot \prod_{q=1}^{s} \left(1 + \frac{2^{b_q}}{n + T_{q-1}}\right).$$

Thus, if for $1 \le p \le r$, $1 \le q \le s$, defining $m_p = \frac{n+S_{p+1}+T_s}{2^{a_p}}$, $m_{r+q} = \frac{n+T_{q-1}}{2^{b_q}}$, then we get the required equality.

We need only to prove that all m_i are integers. For $1 \le p \le r$. we know that

$$n + S_{p+1} + T_s \equiv n + T_s \equiv 0 \pmod{2^{a_p}}.$$

And for $1 \le q \le s$, we have

$$n + T_{q-1} \equiv n + T_s \equiv 0 \pmod{2^{b_q}},$$

which implies the conclusion.

Proof 3. For any $a(\ne 0, -1)$,

$$\left(1 + \frac{1}{a}\right)\left(1 + \frac{1}{a+1}\right) = \left(1 + \frac{1}{\frac{a}{2}}\right). \tag{2}$$

We rewrite the left-hand side of (1) in the form of the product of $2^k - 1$ fractions:

$$\frac{n + 2^k - 1}{n} = \frac{n+1}{n} \cdot \frac{n+2}{n+1} \cdot \frac{n+3}{n+2} \cdots \frac{n + 2^k - 2}{n + 2^k - 3} \cdot \frac{n + 2^k - 1}{n + 2^k - 2}. \tag{3}$$

For even n, grouping successively the right-hand side of (3) in pairs from the left to the right and by using (2), we get the form of product of 2^{k-1} fractions:

$$\left(\frac{n/2+1}{n/2} \cdot \frac{n/2+2}{n/2+1} \cdots \frac{n/2 + 2^{k-1} - 1}{n/2 + 2^{k-1} - 2}\right) \frac{n + 2^k - 1}{n + 2^k - 2}. \tag{4}$$

For odd n, grouping the right-hand side of (3) in pairs from the right to the left and by using (2), we get the form of product of 2^{k-1} fractions:

$$\frac{n+1}{n}\left(\frac{(n+1)/2+1}{(n+1)/2}\cdot\frac{(n+1)/2+2}{(n+1)/2+1}\cdot\right.$$
$$\left....\frac{(n+1)/2+2^{k-1}-2}{(n+1)/2+2^{k-1}-3}\cdot\frac{(n+1)/2+2^{k-1}-1}{(n+1)/2+2^{k-1}-2}\right). \qquad (5)$$

Repeating the above grouping process to the fractions in big parentheses of (4) and (5) $k-2$ times, respectively, we get the form of the right hand side of (1).

【Score Situation】 This particular problem saw the following distribution of scores among contestants: 276 contestants scored 7 points, 5 contestants scored 6 points, 3 contestants scored 5 points, 14 contestants scored 4 points, 6 contestants scored 3 points, 9 contestants scored 2 points, 96 contestants scored 1 point, and 118 contestants scored 0 point. The average score of this problem is 4.108, indicating that it was simple.

Among the top five teams in the team scores, the scores of this problem are as follows: The China team scored 42 points (with a total team score of 208 points), the South Korea team scored 42 points (with a total team score of 204 points), the United States team scored 42 points (with a total team score of 190 points), the Russia team scored 42 points (with a total team score of 187 points), and the North Korea team scored 42 points (with a total team score of 184 points).

The gold medal cutoff for this IMO was set at 31 points (with 45 contestants earning gold medals), the silver medal cutoff was 24 points (with 92 contestants earning silver medals), and the bronze medal cutoff was 15 points (with 141 contestants earning bronze medals).

In this IMO, no contestant achieved a perfect score of 42 points.

3.3 Summary

Overall, there are not many indeterminate equation problems in IMOs, but these problems have never deviated from IMOs, on the other hand. Three indeterminate equation problems have appeared since the 50th IMO, and the frequency is still relatively stable.

In the first 64 IMOs, there were a total of 14 indeterminate equation problems. These problems can be broadly categorized into two types, as depicted in Figure 3.1. The score details for these problems are presented in Table 3.2. Due to the smaller number of participating teams and missing contestant score information in early IMOs, there are several blanks in Table 3.2.

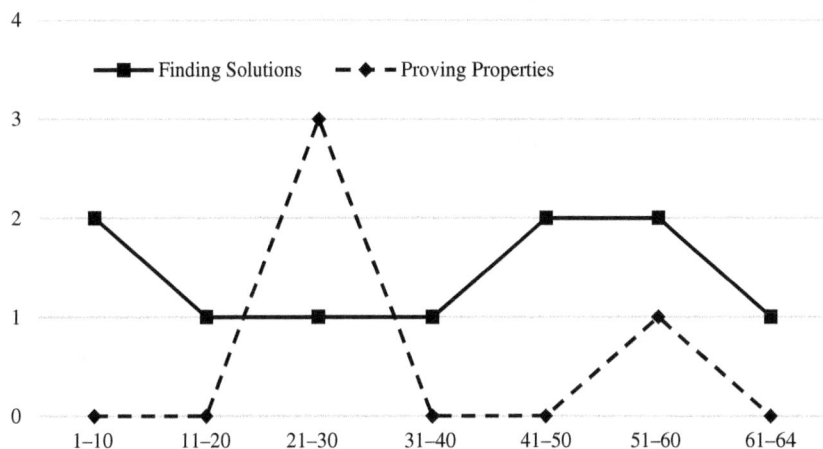

Figure 3.1 Numbers of Indeterminate Equation Problems in the First 64 IMOs

Table 3.2 Score Details of Indeterminate Equation Problems in the First 64 IMOs

Problem	3.1	3.2	3.3	3.4	3.5	3.6	3.7	3.8
Average score	3.800	5.412	3.459	5.216	3.354	2.304	4.998	1.359
Top five mean				5.125	6.833	6.000	6.933	4.300

Problem	3.9	3.10	3.11	3.12	3.13	3.14	
Average score	3.736	3.520	3.782	3.833	0.634	4.108	
Top five mean	6.967	6.733	5.550	4.900	2.033	7.000	

Note. Top five Mean = Total score of the top five teams/Total number of contestants from the top five teams.

Problems 3.1–3.10 focus on "finding solutions of indeterminate equations"; Problems 3.11–3.14 deal with "proving properties satisfied by indeterminate equations".

These 14 problems were proposed by 12 countries. Bulgaria and Germany each contributed two problems.

From Table 3.2, it can be observed that in the first 64 IMOs, there was one indeterminate equation problem with an average score of 0–1 points; one problem with an average score of 1–2 points; one problem with an average score of 2–3 points; seven problems with an average score of 3–4 points; four problems with an average score above 4 points. Overall, the indeterminate equation problems were relatively straightforward.

Among these 14 problems, the problem with the lowest average score is Problem 3.13 (IMO 29-6, contributed by Germany, proving properties satisfied by indeterminate equations).

Appendix A

IMO General Information

Session	Year	Host	Number of Participating Teams	Number of Contestants	Gold (Cutoffs/Numbers of Medalists)	Silver (Cutoffs/Numbers of Medalists)	Bronze (Cutoffs/Numbers of Medalists)
IMO 1	1959	Romania	7	52	37/3	36/3	33/5
IMO 2	1960	Romania	5	39	40/4	37/4	33/4
IMO 3	1961	Hungary	6	48	37/3	34/4	30/4
IMO 4	1962	Czechoslovakia	7	56	41/4	34/12	29/15
IMO 5	1963	Poland	8	64	35/7	28/11	21/17
IMO 6	1964	The Union of Soviet Socialist Republics	9	72	38/7	31/9	27/19
IMO 7	1965	The German Democratic Republic	10	80	38/8	30/12	20/17
IMO 8	1966	Bulgaria	9	72	39/13	34/15	31/11
IMO 9	1967	Yugoslavia	13	99	38/11	30/14	22/26
IMO 10	1968	The Union of Soviet Socialist Republics	12	96	39/22	33/22	26/20
IMO 11	1969	Romania	14	112	40/3	30/20	24/21
IMO 12	1970	Hungary	14	112	37/7	30/11	19/40

(*Continued*)

(*Continued*)

Session	Year	Host	Number of Participating Teams	Number of Contestants	Gold	Silver	Bronze
					(Cutoffs/Numbers of Medalists)		
IMO 13	1971	Czechoslovakia	15	115	35/7	23/12	11/29
IMO 14	1972	Poland	14	107	40/8	30/16	19/30
IMO 15	1973	The Union of Soviet Socialist Republics	16	125	35/5	27/15	17/48
IMO 16	1974	The German Democratic Republic	18	140	38/10	30/24	23/37
IMO 17	1975	Bulgaria	17	135	38/8	32/25	23/36
IMO 18	1976	Austria	18	139	34/9	23/28	15/45
IMO 19	1977	Yugoslavia	21	155	34/13	24/29	17/35
IMO 20	1978	Romania	17	132	35/5	27/20	22/38
IMO 21	1979	The United Kingdom	23	166	37/8	29/32	20/42
IMO 22	1981	The United States of America	27	185	41/36	34/37	26/30
IMO 23	1982	Hungary	30	119	37/10	30/20	21/31
IMO 24	1983	France	32	186	38/9	26/27	15/57
IMO 25	1984	Czechoslovakia	34	192	40/14	26/35	17/49
IMO 26	1985	Finland	38	209	34/14	22/35	15/52
IMO 27	1986	Poland	37	210	34/18	26/41	17/48
IMO 28	1987	Cuba	42	237	42/22	32/42	18/56
IMO 29	1988	Australia	49	268	32/17	23/48	14/65
IMO 30	1989	Germany	50	291	38/20	30/55	18/72
IMO 31	1990	The People's Republic of China	54	308	34/23	23/56	16/76
IMO 32	1991	Sweden	56	318	39/20	31/51	19/84
IMO 33	1992	The Russian Federation	56	322	32/26	24/55	14/74

(*Continued*)

Session	Year	Host	Number of Participating Teams	Number of Contestants	Gold	Silver	Bronze
					\(\dfrac{\text{Cutoffs/Numbers}}{\text{of Medalists}}\)		
IMO 34	1993	Turkey	73	413	30/35	20/66	11/97
IMO 35	1994	Chinese Hong Kong	69	385	40/30	30/64	19/98
IMO 36	1995	Canada	73	412	37/30	29/71	19/100
IMO 37	1996	India	75	424	28/35	20/66	12/99
IMO 38	1997	Argentina	82	460	35/39	25/70	15/122
IMO 39	1998	Chinese Taiwan	76	419	31/37	24/66	14/102
IMO 40	1999	Romania	81	450	28/38	19/70	12/118
IMO 41	2000	Republic of Korea	82	461	30/39	21/71	11/119
IMO 42	2001	The United States of America	83	473	30/39	20/81	11/122
IMO 43	2002	The United Kingdom	84	479	29/39	23/73	14/120
IMO 44	2003	Japan	82	457	29/37	19/69	13/104
IMO 45	2004	Greece	85	486	32/45	24/78	16/120
IMO 46	2005	Mexico	91	513	35/42	23/79	12/128
IMO 47	2006	Slovenia	90	498	28/42	19/89	15/122
IMO 48	2007	Vietnam	93	520	29/39	21/83	14/131
IMO 49	2008	Spain	97	535	31/47	22/100	15/120
IMO 50	2009	Germany	104	565	32/49	24/98	14/135
IMO 51	2010	Kazakhstan	95	522	27/47	21/103	15/115
IMO 52	2011	The Netherlands	101	563	28/54	22/90	16/137
IMO 53	2012	Argentina	100	547	28/51	21/88	14/137
IMO 54	2013	Colombia	97	527	31/45	24/92	15/141
IMO 55	2014	South Africa	101	560	29/49	22/113	16/133
IMO 56	2015	Thailand	104	577	26/39	19/100	14/143
IMO 57	2016	Chinese Hong Kong	109	602	29/44	22/101	16/135

(*Continued*)

(*Continued*)

Session	Year	Host	Number of Participating Teams	Number of Contestants	Gold	Silver	Bronze
					(Cutoffs/Numbers of Medalists)		
IMO 58	2017	Brazil	111	615	25/48	19/90	16/153
IMO 59	2018	Romania	107	594	31/48	25/98	16/143
IMO 60	2019	The United Kingdom	112	621	31/52	24/94	17/156
IMO 61	2020	The Russian Federation	105	616	31/49	24/112	15/155
IMO 62	2021	The Russian Federation	107	619	24/52	19/103	12/148
IMO 63	2022	Norway	104	589	34/44	29/101	23/140
IMO 64	2023	Japan	112	618	32/54	25/90	18/170

Appendix B

IMO Number Theory Problem Index

Problem Number in the IMO	Proposing Country	Category	Problem Number in the Book	Page Number
1-1	Poland	Discussions on Divisibility, Divisibility of Integers	Problem 1.1	31
2-1	Bulgaria	Finding Solutions, Indeterminate Equations	Problem 3.1	156
4-1	Poland	Finding Solutions, Indeterminate Equations	Problem 3.2	157
6-1	Czechoslovakia	Existence Problems, Modular Arithmetic	Problem 2.1	99
9-3	United Kingdom	Discussions on Divisibility, Divisibility of Integers	Problem 1.2	32
10-2	Czechoslovakia	Other Problems, Divisibility of Integers	Problem 1.29	75
10-6	The United Kingdom	Other Problems, Divisibility of Integers	Problem 1.30	76
11-1	The German Democratic Republic	Primes, Divisibility of Integers	Problem 1.16	49

(Continued)

(*Continued*)

Problem Number in the IMO	Proposing Country	Category	Problem Number in the Book	Page Number
12-2	Romania	Other Problems, Divisibility of Integers	Problem 1.31	78
13-3	Poland	Exploring Relationships, Modular Arithmetic	Problem 2.19	130
14-3	The United Kingdom	Discussions on Divisibility, Divisibility of Integers	Problem 1.3	33
16-3	Romania	Existence Problems, Modular Arithmetic	Problem 2.2	100
17-2	The United Kingdom	Existence Problems, Modular Arithmetic	Problem 2.3	101
17-4	The Soviet Union	Finding Numbers, Modular Arithmetic	Problem 2.11	116
19-3	The Netherlands	Discussions on Divisibility, Divisibility of Integers	Problem 1.4	35
19-5	The German Democratic Republic	Finding Solutions, Indeterminate Equations	Problem 3.3	158
20-1	Cuba	Maximum or Minimum Values, Modular Arithmetic	Problem 2.23	137
21-1	Germany	Discussions on Divisibility, Divisibility of Integers	Problem 1.5	36
22-3	The Netherlands	Finding Solutions, Indeterminate Equations	Problem 3.4	160

(*Continued*)

(*Continued*)

Problem Number in the IMO	Proposing Country	Category	Problem Number in the Book	Page Number
22-4	Belgium	Discussions on Divisibility, Divisibility of Integers	Problem 1.6	37
23-4	The United Kingdom	Proving Properties, Indeterminate Equations	Problem 3.11	171
24-3	Germany	Primes, Divisibility of Integers	Problem 1.17	51
25-2	The Netherlands	Discussions on Divisibility, Divisibility of Integers	Problem 1.7	38
25-6	Poland	Discussions on Divisibility, Divisibility of Integers	Problem 1.8	39
26-2	Australia	Exploring Relationships, Modular Arithmetic	Problem 2.20	132
27-1	Germany	Proving Properties, Indeterminate Equations	Problem 3.12	173
28-6	The Soviet Union	Existence Problems, Modular Arithmetic	Problem 2.4	102
29-6	Germany	Proving Properties, Indeterminate Equations	Problem 3.13	174
30-5	Sweden	Existence Problems, Modular Arithmetic	Problem 2.5	104
31-3	Romania	Finding Numbers, Modular Arithmetic	Problem 2.12	117
32-2	Romania	Primes, Divisibility of Integers	Problem 1.18	53

(*Continued*)

(*Continued*)

Problem Number in the IMO	Proposing Country	Category	Problem Number in the Book	Page Number
33-1	New Zealand	Discussions on Divisibility, Divisibility of Integers	Problem 1.9	40
33-6	The United Kingdom	Other Problems, Divisibility of Integers	Problem 1.32	79
35-3	Romania	Other Problems, Divisibility of Integers	Problem 1.33	81
35-4	Australia	Discussions on Divisibility, Divisibility of Integers	Problem 1.10	42
35-6	Finland	Primes, Divisibility of Integers	Problem 1.19	55
36-6	Poland	Finding Numbers, Modular Arithmetic	Problem 2.13	119
37-3	Romania	Function Related Problems, Divisibility of Integers	Problem 1.24	65
37-4	Russia	Maximum or Minimum Values, Modular Arithmetic	Problem 2.24	138
38-5	Czech Republic	Finding Solutions, Indeterminate Equations	Problem 3.5	161
38-6	Lithuania	Function Related Problems, Divisibility of Integers	Problem 1.25	67
39-3	Byelorussia	Primes, Divisibility of Integers	Problem 1.20	56
39-4	The United Kingdom	Discussions on Divisibility, Divisibility of Integers	Problem 1.11	43
39-6	Bulgaria	Function Related Problems, Divisibility of Integers	Problem 1.26	69

(*Continued*)

(Continued)

Problem Number in the IMO	Proposing Country	Category	Problem Number in the Book	Page Number
40-4	Chinese Taiwan	Finding Numbers, Modular Arithmetic	Problem 2.14	120
41-5	Russia	Existence Problems, Modular Arithmetic	Problem 2.6	105
42-4	Canada	Exploring Relationships, Modular Arithmetic	Problem 2.21	133
42-6	Bulgaria	Primes, Divisibility of Integers	Problem 1.21	58
43-3	Romania	Discussions on Divisibility, Divisibility of Integers	Problem 1.12	44
43-4	Romania	Discussions on Divisibility, Divisibility of Integers	Problem 1.13	46
44-2	Bulgaria	Finding Solutions, Indeterminate Equations	Problem 3.6	162
44-6	France	Existence Problems, Modular Arithmetic	Problem 2.7	107
45-6	Iran	Finding Numbers, Modular Arithmetic	Problem 2.15	122
46-2	The Netherlands	Exploring Relationships, Modular Arithmetic	Problem 2.22	135
46-4	Poland	Finding Numbers, Modular Arithmetic	Problem 2.16	125
47-4	The United States	Finding Solutions, Indeterminate Equations	Problem 3.7	165
48-5	The United Kingdom	Finding Numbers, Modular Arithmetic	Problem 2.17	127

(Continued)

(*Continued*)

Problem Number in the IMO	Proposing Country	Category	Problem Number in the Book	Page Number
49-3	Lithuania	Existence Problems, Modular Arithmetic	Problem 2.8	108
50-1	Australia	Discussions on Divisibility, Divisibility of Integers	Problem 1.14	47
51-3	The United States	Function Related Problems, Divisibility of Integers	Problem 1.27	72
52-5	Iran	Function Related Problems, Divisibility of Integers	Problem 1.28	74
53-6	Serbia	Finding Numbers, Modular Arithmetic	Problem 2.18	128
54-1	Japan	Proving Properties, Indeterminate Equations	Problem 3.14	175
55-5	Luxembourg	Other Problems, Divisibility of Integers	Problem 1.34	83
56-2	Serbia	Finding Solutions, Indeterminate Equations	Problem 3.8	166
57-3	Russia	Other Problems, Divisibility of Integers	Problem 1.35	84
57-4	Luxembourg	Maximum or Minimum Values, Modular Arithmetic	Problem 2.25	139
58-6	The United States	Existence Problems, Modular Arithmetic	Problem 2.9	110
59-5	Mongolia	Primes, Divisibility of Integers	Problem 1.22	61
60-4	El Salvador	Finding Solutions, Indeterminate Equations	Problem 3.9	169

(*Continued*)

(*Continued*)

Problem Number in the IMO	Proposing Country	Category	Problem Number in the Book	Page Number
61-5	Estonia	Primes, Divisibility of Integers	Problem 1.23	63
62-1	Australia	Other Problems, Divisibility of Integers	Problem 1.36	86
63-3	The United States	Existence Problems, Modular Arithmetic	Problem 2.10	113
63-5	Belgium	Finding Solutions, Indeterminate Equations	Problem 3.10	170
64-1	Colombia	Discussions on Divisibility, Divisibility of Integers	Problem 1.15	48

www.ingramcontent.com/pod-product-compliance
Lightning Source LLC
Chambersburg PA
CBHW061250220326
41599CB00028B/5598